A Primer in Frustrated Lewis Pair Hydrogenation

Concepts to Applications

A Primer in Frustrated Lewis Pair Hydrogenation

Concepts to Applications

By

Douglas W. Stephan

University of Toronto, Canada
Email: dstephan@chem.utoronto.ca

ROYAL SOCIETY
OF **CHEMISTRY**

Print ISBN: 978-1-83916-244-2
EPUB ISBN: 978-1-83916-506-1

A catalogue record for this book is available from the British Library

The Royal Society of Chemistry is a charity, registered in England and Wales, Number 207890, and a company incorporated in England by Royal Charter (Registered No. RC000524), registered office: Burlington House, Piccadilly, London W1J 0BA, UK, Telephone: +44 (0) 20 7437 8656.

Visit our website at www.rsc.org/books

Printed in the United Kingdom by CPI Group (UK) Ltd, Croydon, CR0 4YY, UK

Dedication

In 1980, ten years after our first date, I acknowledged Dianne Stephan in my PhD thesis, thanking her for "being herself and letting me be me". Now, over 40 years later, it is even more appropriate that I dedicate this text to her. Dianne has always been there, unendingly flexible, tolerant of my obsession with chemistry, supportive of my ambitions, and continually the love of my life.

A Primer in Frustrated Lewis Pair Hydrogenation: Concepts to Applications
By Douglas W. Stephan
© Douglas W. Stephan 2022
Published by the Royal Society of Chemistry, www.rsc.org

Preface

Over 45 years ago, while performing my senior undergraduate-research project at McMaster University, my supervisor and I were reviewing spectral data on a reaction I had attempted. After some thought, Professor McGlinchey concluded I had made a new compound. He then told me I was the first person to have ever made this species. Those words were exhilarating, and the thrill of discovery stuck with me and has been a motivating force ever since.

Some years later, as a professor myself, I had an active research group in organometallic chemistry. We had spent about 20 years studying early transition metal chemistry, probing early–late heterobimetallic systems, titanium–sulfur and zirconium–phosphorus chemistry. We had also uncovered that previously little used ligands, phosphinimides, provided uniquely reactive early metal olefin polymerization catalysts. It was this later finding that had prompted a very fruitful collaboration with NOVA Chemicals. This collaboration led us to use the Lewis acid, $B(C_6F_5)_3$, as this was firmly established in the late 1990s as an activator for early-transition metal polymerization catalysts. In attempting to generate phosphine-stabilized titanium compounds, one of my MSc students, Lourisa Cabreara, found an unexpected product that contained no titanium when the reactions were done with bulky phosphines. In these cases, the phosphines effected *para*-attack on the borane affording the zwitterion $R_3PC_6F_4BF(C_6F_5)_2$. Shortly after this work, Greg Welch joined the group as a new PhD student. In suggesting a project, I proposed that he repeat Lourisa's reactions with bulky secondary phosphines, thinking this would be an avenue to unusual anionic phosphines, an underexplored class of ligands in transition metal chemistry. While this was the original motivation, it was curiosity around the fundamental reactivity that led us to the discovery of the first main group system, $R_2PHC_6F_4BH(C_6F_5)_2$, that could reversibly activate dihydrogen.

With that finding, the stage was set for a renaissance of main group chemistry. The emergence of the initially simple notion of 'frustrated Lewis pairs (FLPs)' has certainly evolved and broadened over the past 15 years and led to numerous new protocols useful in synthetic chemistry. It is also clear that the concept of FLPs provide a new axiom of chemical reactivity, one that is not limited to classical main group Lewis acids and bases.

A Primer in Frustrated Lewis Pair Hydrogenation: Concepts to Applications
By Douglas W. Stephan
© Douglas W. Stephan 2022
Published by the Royal Society of Chemistry, www.rsc.org

This book focuses on the development of FLP chemistry involving dihydrogen, from the initial discovery to the development of applications and beyond.

I would like to extend my most sincere thanks to the undergraduates, graduate students, postdoctoral fellows, and colleagues who have worked so creatively on the development of frustrated Lewis pair chemistry. This large group of coworkers, colleagues and friends have shared the joy of discovery and I cannot thank them enough. I also wish to thank the current group members who assisted with the editing of this manuscript.

I am also grateful to the funding agencies that have supported our efforts on FLP chemistry, in particular I would like to offer special thanks to the NSERC of Canada for continuous support for over 40 years as well as the award of Canada research Chairs at the University of Windsor (2005–2007) and the University of Toronto (2008–2022) and the 2019 Polanyi Prize. In addition, the author is grateful to the Humboldt (2003) and Killam (2009–2011) foundations for fellowships that have allowed the author the time to focus on the development of this area of chemistry. The Einstein Foundation is acknowledged for the Visiting Professorship at TU-Berlin (2016–2019). The Guggenheim foundation in the USA is thanked for the award of a 2020 Fellowship and the Canada Council for the Arts is thanked for the award of the 2021 Killam Prize in Natural Science.

Douglas W. Stephan
Department of Chemistry, University of Toronto, Canada

Contents

A Primer in Frustrated Lewis Pair Hydrogenation: Concepts to Applications
By Douglas W. Stephan
© Douglas W. Stephan 2022
Published by the Royal Society of Chemistry, www.rsc.org

1 Context, Background, and Discovery

1.1 Chapter Overview

To begin a discussion of 'frustrated Lewis pair (FLP) chemistry', we set the stage as it was at the turn of the 20th to 21st century. At that time, it was clear that the implementations of catalytic processes had been a major driver in the evolution of chemistry and consequently in many technological advances in society. Herein, we briefly describe the history of this important chemical concept beginning with its inception over 100 years ago. We discuss why transition metals are so well suited to catalysis, focusing on their unique ability to act concurrently as both electron donors and acceptors. This is contrasted with the perceptions of main group elements, where generally molecules containing these elements act as either Lewis acids or Lewis bases. We also discuss examples of chemical reactivity that contravene reaction pathways expected based on the simple concepts of Lewis acidity and basicity. These examples prompted us to consider the possibility that combinations of Lewis acids and bases could mimic the reactivity of transition metals. This led to the initial unveiling of the notion of 'frustrated Lewis pairs (FLPs)' in which select combinations of main group donors and acceptors could activate dihydrogen in addition to other small molecules.

1.2 Our Chemical World

While it is shocking to compare our lives today with that of our parents, grandparents, or great grandparents, there is no doubt that over the last century, technological advances have dramatically transformed the way we live. Modern life, in both work and play, is replete with the tools, products, and materials derived from science and engineering that have either eased or enriched our lives. What is perhaps less widely appreciated, is the role that fundamental chemistry plays in cultivating the development and commercialization of such technologies. Indeed, it is not an overstatement to say that the nature of modern society has been shaped in large part by activities initiated in

A Primer in Frustrated Lewis Pair Hydrogenation: Concepts to Applications
By Douglas W. Stephan
© Douglas W. Stephan 2022
Published by the Royal Society of Chemistry, www.rsc.org

research laboratories around the world. Life-saving drugs, synthetic materials, plastics, solar energy, the internet, computers, and cell phones are just a few important examples of the endless array of modern technologies where chemistry is at the root of these advances, in whole or in part. Drugs, agrochemicals, and polymers have obvious roots in fundamental organic chemistry. For other technological advances, such as the ability to miniaturize circuitry, the precise control of the properties of new materials, the generation of vibrant colors on a display or the ability to harvest solar energy, the link to chemistry is perhaps less obvious to the uninitiated. Nonetheless, such advances are derived from an understanding of chemical properties and how to efficiently make and alter compounds or materials for a specific purpose. Ultimately, the recipe for cutting-edge technologies has involved a mix of fundamental chemistry, innovative engineering, and creative entrepreneurship.

1.3 A Brief History of Catalysis

Catalysis is an advance in chemistry that has contributed disproportionately to our abilities to synthesize new products over the past century. The origin of this transformative concept traces back to Ostwald (Figure 1.1), who was awarded the 1909 Nobel Prize for conceiving the notion of 'catalytic phenomena as a means to accelerate chemical processes'. Subsequent work by Sabatier[1,2] (1912 Nobel Prize, shared with Grignard) was similarly rewarded three years later. In his seminal findings, Sabatier experimentally showed that amorphous transition metals, principally nickel, can be used to catalyze the hydrogenation of unsaturated organic molecules. This finding provided a powerful method for efficient access to a range of organic compounds and initiated the field of heterogeneous catalysis. In 1918, Haber[3] was recognized for the catalytic production of ammonia, the precursor to fertilizer. This discovery has been described as the most important of the 20th century, as the Haber process provides the capability to feed the world's ever-growing population.

Fifty years later, Ziegler[4] and Natta[5–8] (1963 Nobel Prize) uncovered heterogeneous catalysts for the polymerization of ethylene and propylene to the respective polymers. While these findings ushered in the industrial production of plastics, this field has since broadened, providing a range of synthetic polymers that have found applications in a wide array of consumer products.

Concept	Hydrogenation Catalysis	NH$_3$ Synthesis	Polymerization	Organometallics	Asymmetric catalysis	Metathesis	Cross-coupling
1909	1912	1918	1963	1973	2001	2005	2010
Ostwald	Sabatier	Haber	Ziegler	Wilkinson	Knowles	Grubbs	Heck
			Natta	Fischer	Noyori	Schrock	Negishi
					Sharpless	Chauvin	Suzuki

Figure 1.1   Nobel Laureates in chemical catalysis.

In the 1960s, the emergence of organometallic chemistry brought molecular chemistry to catalysis. In this regard, a classical example is the work of Wilkinson (1973 Nobel Prize, shared with Fischer) and coworkers. These researchers showed that the rhodium complex, $(Ph_3P)_3RhCl$, acts as a pre-catalyst for the hydrogenation of unsaturated organic compounds in solution under mild conditions.[9,10] This finding was seminal as it allowed chemists to precisely define and tune the molecular environment about a transition metal center for utilization in specific and desirable chemical transformations. Further advances within organometallic chemistry have allowed the development of powerful new synthetic protocols. One such example is the design and development of homogeneous catalysts for the enantioselective synthesis of many organic products. These advances have been widely exploited for drug and agrochemical products and were recognized with the 2001 Nobel prize (Knowles,[11] Noyori[12] and Sharpless[13]). A finding of similarly broad impact has been the advent of transition metal catalysts for olefin metathesis. This strategy for the rearrangement of C=C double bonds has proved tremendously versatile for the construction of new organic pharmaceuticals, polymers, and materials. The impact of this development was recognized with the 2005 Nobel Prize to Grubbs,[14] Schrock[15] and Chauvin.[16] Finally, metal-mediated cross-coupling catalysis is another powerful tool for C–C bond formation. Indeed, over the past 30 years, these methodologies have become classic strategies in synthetic organic chemistry and thus are widely employed in academia and industry, prompting the 2010 Nobel Prize to Negishi,[17] Suzuki[18] and Heck.[19]

Beyond the above-mentioned seminal, Nobel-garnering, discoveries, numerous other advances in heterogeneous and homogeneous catalysis have re-defined the synthetic toolbox, providing a myriad of transition metal-mediated strategies for chemical bond construction. Many of these technologies have been exploited to generate or redefine chemical industries providing new protocols and ready access to new and useful products. Indeed, it is estimated that 85% of all products we use or consume today required at least one step involving catalysis.[20]

1.4   Transition Metal Activation of Small Molecules

Having established the broad utility of transition metal in catalysis, we now examine the reasons behind this. Ultimately, the ability of transition metals to adopt varying oxidation states and to activate stable small molecules is fundamental to their applications in catalysis. To probe the latter feature more deeply, we now consider two quintessential examples involving the interaction of metal centers with dihydrogen and olefins. These will serve to illustrate the core features of transition metals that facilitate catalytic reactions.

The interaction of dihydrogen with a metal involves the donation of electron density from the H–H σ-bond to a vacant metal orbital (Figure 1.2). Back donation from the filled metal d-orbital to the σ*-orbital of dihydrogen results in the activation of the H–H bond typically yielding a metal-dihydride species. Of course, should the degree of back donation be insufficient to fully populate the σ*-orbital, the H–H bond is not cleaved and a dihydrogen complex,[21,22] in which the metal is not formally oxidized, is the result.

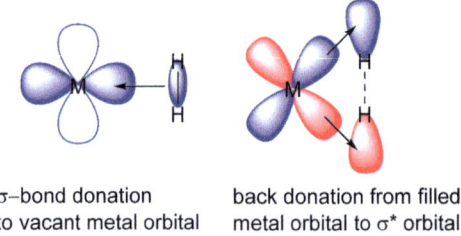

σ–bond donation
to vacant metal orbital

back donation from filled
metal orbital to σ* orbital

Figure 1.2 Transition metal interactions with dihydrogen.

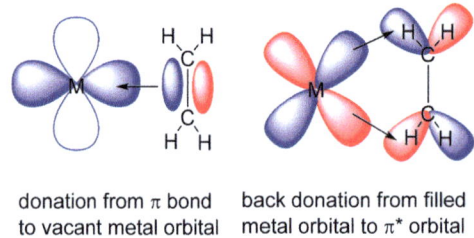

donation from π bond
to vacant metal orbital

back donation from filled
metal orbital to π* orbital

Figure 1.3 Transition metal interactions with ethylene.

However, in catalytic hydrogenation, repetitive cleavage of the H–H bond is required and this hinges on the ability of the metal to act concurrently as an electron acceptor and an electron donor.

In the case of a transition metal interacting with ethylene, the Dewar[23]–Chatt–Duncanson[24] description provides an understanding of the bonding. The first interaction involves the donation from the π-system of ethylene to a vacant d-orbital on the transition metal center providing a σ-interaction (Figure 1.3). Concurrent back-donation from the filled metal d-orbitals to the π*-orbital of ethylene results in stronger metal–olefin binding and diminished C–C bond strength. Thus, in these σ- and π-interactions, the metal acts as both an electron acceptor and as an electron donor, affecting the partial reduction of the olefinic bond.

The above examples suffice to illustrate the key aspect involved in the interactions of transition metals with small molecules, namely the ability to act simultaneously as electron acceptors and electron donors. These properties of transition metal complexes can be tuned *via* ligand modification, which allows for the optimization of the metal species for the activation of other substrates, well beyond dihydrogen and olefins. It is this strategy that has afforded the myriad of metal-mediated catalytic transformations. The broad success in the development of transition metal catalysts that had occurred by the turn of the 20th century led to the common belief that only transition metals had the correct balance of acceptor and donor properties suitable for catalysis involving small molecule activation.

1.5   Transition Metals: the Downside

Despite the impact transition metal-mediated processes have had on synthetic chemistry, the use of transition metal-based catalysts also has disadvantages. The transition metals most often used for catalysis are derived from heavy, precious metals such as Pt, Rh, Ru, Ir, and Pd. These metals are generally quite toxic to humans. This means that considerable effort and cost must be undertaken to remove metallic residues from products meant for human consumption. Indeed, the daily exposure limits set by the US-FDA (Table 1.1) illustrate that most metals used in catalysis (Pd, Rh, Ru, and Pt) are about 100 µg per day. Interestingly, while these limits are higher than that of more commonly known toxic elements such as lead or arsenic, the tolerance of the precious metals is substantially lower than that of tin (Table 1.1).

Precious metals are expectedly expensive (Table 1.1), ranging from $24–6000 per troy oz at this writing. Moreover, as is evident from their collective name, many of these elements are in low abundance on earth (Table 1.1). The consequence is a considerable carbon footprint as mining, refining, and recycling of these metals is energy intensive. These problematic features have motivated recent efforts in the organometallic and catalysis communities, prompting the development of catalysts derived from more earth-abundant, less toxic, first-row transition metals.[25,26] These advances aside, are there other avenues?

1.6   Main Group Elements

Conceptually, another approach to avoid the cost and toxicity of the precious metals could be the use of p-block elements. Indeed, these elements are typically in significantly higher abundance on earth, and much cheaper than the precious metals (Table 1.2). Nonetheless, applications of these elements in industrial chemical production have been few and far between. Certainly, phosphorus compounds such as phosphines have been widely exploited as 'by-stander' donor ligands for transition metal catalysts. Such ligands can be readily modified to tune the reactivity at the metal center.

Table 1.1   Permitted daily exposure, cost, and natural abundance for selected transition metal elements.

Element	Permitted exposure/µg per day	Cost/$ oz^{-1}	Abundance/ ppm	Element	Permitted exposure/ µg per day	Cost/$ oz^{-1}	Abundance/ ppm
Pb	5		14	Os	100	400	0.0015
As	15		1.8	Rh	100	6000	0.001
Hg	30		0.085	Ru	100	244	0.001
Au	100	1902	0.004	Ag	150	24	0.075
Pd	100	876	0.015	Pt	100	1230	0.005
Ir	100	1670	0.001	Sn	6000		2.3

Table 1.2   Cost and natural abundance for selected main group elements.

Element	Cost/\$ kg^{-1}	Abundance/ppm	Element	Cost/\$ kg^{-1}	Abundance/ppm
B	3.7	10	N	0.14	19
Al	1.8	82 300	P	2.7	1050
Ga	148	19	As	1.2	1.8
C	0.12	200	O	0.15	461 000
Si	1.7	282 000	S	0.09	350
Ge	914	1.5	Se	21.4	0.05

Thus, while the nature of the main group species is important in determining the reactivity, it is ancillary to the site of principle reactivity at the metal. In a similar sense, the main group aluminium compounds, known as alanes, are employed as activators in the polymerization of olefins. Here, the main group element plays an ancillary role, serving to both alkylate and abstract a substituent from a group IV metal. The ability of aluminium to act as an initiator thus provides a highly reactive metal center where the polymerization of olefins is catalyzed.

The above examples of main group reagents in catalysis serve to demonstrate general features that characterized the reactivity of main group elements in the late 20th century. In contrast to metal centers, main group compounds were not generally known for their ability to change oxidation state. Rather, the reactivity of main group elements was typically confined to the ability to act as either electron donors or electron acceptors.

1.7   Lewis Acids and Bases

It was the seminal work of Gilbert Lewis[27] in the 1920s that drew the analogy of the chemistry of main group elements to that of Brønsted acids and bases, describing electron acceptors as acids and electron donors as bases. Thus, while neutralization occurs upon the combination of a Brønsted acid and base, yielding water, the combination of a Lewis acid and Lewis base are neutralized by the formation of what is described as a 'classical Lewis acid–base adduct'. A prototypical example of this reactivity is the combination of BF_3 and NH_3. The former molecule contains a formally three-coordinate, electron-deficient boron center that acts as an electron acceptor, whereas the central atom in NH_3 is a pyramidal nitrogen atom with a lone pair of electrons formally occupying one of the sp^3 hybrid orbitals. The combination of these reagents results in the immediate quenching of the Lewis acidity and basicity with the formation of the B–N bond in the Lewis acid–base adduct, F_3BNH_3 (Scheme 1.1). Putting this concept in more modern terms, a Lewis acid is characterized by a low-lying LUMO, which is primed to interact with the lone electron-pair occupying a relatively high energy HOMO of a Lewis base.[28]

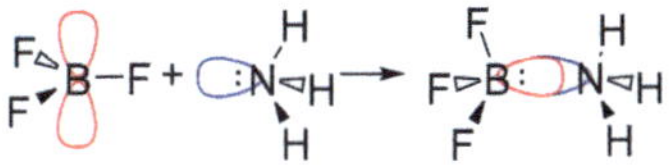

Scheme 1.1 Formation of the Lewis acid-base adduct, F_3BNH_3.

While much of main group chemistry can be described in these terms, the concept is more general than that, as the axioms articulated by Lewis are also pertinent to coordination chemistry.[29,30] Indeed, the interaction of a metal and a ligand can be viewed simply as a Lewis basic donor interacting with a Lewis acidic metal center. The power and elegance of Lewis' empirical ideas lie in its simplicity, while it provides a conceptual understanding of the relationship between valence electron-count and chemical behavior. It is for these reasons that these chemical concepts remain among the first we teach students about chemical reactivity even though these notions were formulated over 100 years ago.

1.8 Contravening Lewis' Rules

While the notion of Lewis acids, bases, and Lewis acid–base adducts accounts for much of main group chemistry, there are reports in the older literature of instances where the chemistry deviates from Lewis' axioms. For example, the 1942 work of H.C. Brown and coworkers[31] examined the interaction of pyridines with simple boranes. They noted that most combinations of Lewis acids and bases, such as lutidine and BF_3, formed the stable classical Lewis adducts. In contrast, lutidine and BMe_3 did not react.[31,32] To attempt to probe this observation, the authors examined molecular models and attributed the absence of interaction in the latter case to the steric conflict of the *ortho*-methyl group of lutidine with the methyl groups on the borane (Figure 1.4).

In 1950, Wittig described the reaction of $Na[CPh_3]$ with $(THF)BPh_3$[33] expecting the conventional Lewis acid–base adduct between the trityl anion and the borane *via* displacement of the weaker Lewis base, THF. However, this was not the case. Instead, this combination led to the ring-opening of THF to afford the anion $[Ph_3C(CH_2)_4OBPh_3]^-$ (Scheme 1.2). A related example emerged in 1959 when Wittig and Benz generated 1,2-dihydrobenzyne *in situ* from *ortho*-fluorobromobenzene and reacted it with a mixture of triphenylphosphine and triphenyl borane to give the *o*-phenylene-linked phosphonium-borate $(C_6H_4)(PPh_3)(BPh_3)$.[34] This reactivity proceeded despite the expectation that the combination of Ph_3P and BPh_3 would form the classical Lewis acid–base adduct Ph_3PBPh_3 (Scheme 1.2). In a related observation a few years later, Wittig and Tochtermann found that $Na[CPh_3]$ and BPh_3 reacted with butadiene to give $Na[Ph_3CCH_2CH(BPh_3)CHCH_2]$ (Scheme 1.2), instead of forming the adduct or effecting anionic polymerization of butadiene. The authors attributed this reactivity to the bulky nature of the Lewis acid/base pairs which precluded classical Lewis acid–base adduct formation. Moreover, this prompted Tochtermann to label such non-quenched Lewis-pairs as 'antagonistisches Paars'.[35]

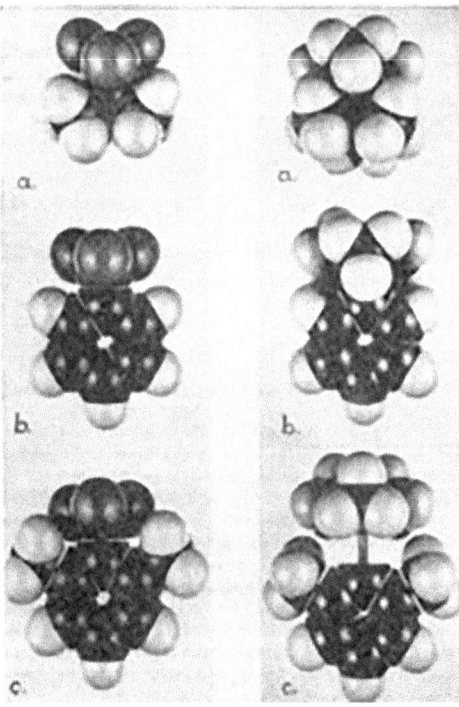

Figure 1.4   Molecular models of reactions of (a) NMe$_3$, (b) pyridine and (c) lutidine interactions with BF$_3$ (left) and BMe$_3$ (right), respectively.[31] Reproduced from ref. 31 with permission from American Chemical Society, Copyright 1942.

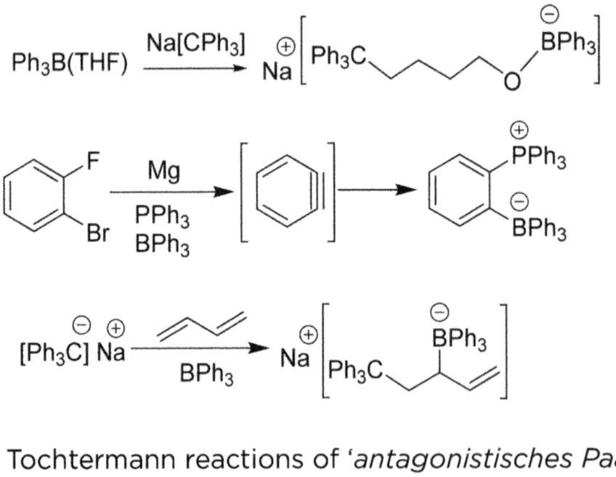

Scheme 1.2   Wittig and Tochtermann reactions of *'antagonistisches Paars'*.

Since these early studies, related THF ring-opening reactions have appeared. In 1992, we described the treatment of ZrCl$_4$(THF)$_2$ with PCy$_3$ to produce the di-zwitterionic dimer [Cl$_4$Zr(μ-O(CH$_2$)$_4$PCy$_3$)]$_2$ (Scheme 1.3).[36] Similar results have been subsequently described for related Lewis acidic metals such as U,[37,38] Sm,[39] Ti,[40] and Zr.[36,41,42] as well as for the combinations of main group Lewis acids with either N or P-based Lewis donors.[43–46]

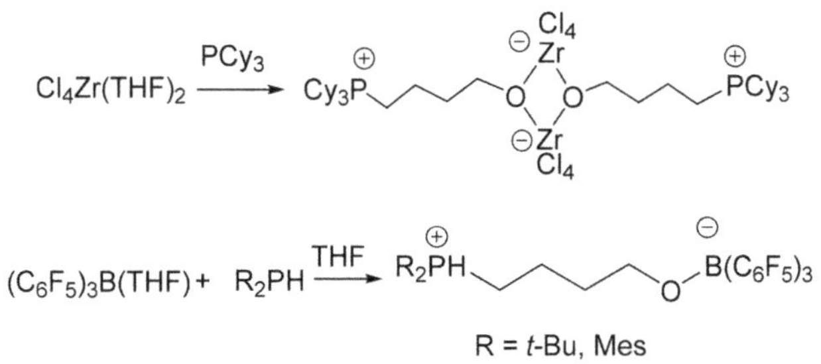

Scheme 1.3 Ring-opening of THF by zirconium and phosphine, Lewis acid-base combinations.

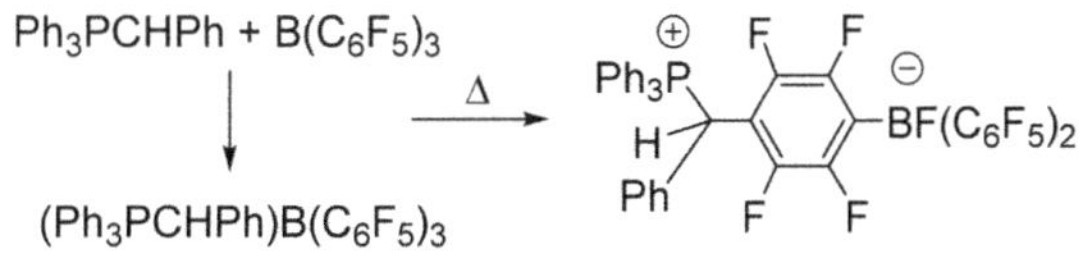

Scheme 1.4 Reactions of the ylide Ph_3PCHPh with $B(C_6F_5)_3$.

In a subsequent study, simple combinations of $(THF)B(C_6F_5)_3$ with phosphines were probed. Again, in most cases, the better electron donor, phosphine would displace THF to give the stronger dative P–B bonded species $(R_3P)B(C_6F_5)_3$. However, sterically encumbered phosphines again resulted in the ring-opening of THF, affording the zwitterionic species of the form $HR_2PC_4H_8OB(C_6F_5)_3$ (R = t-Bu, Mes) (Scheme 1.3).[47] In these cases, rather than the base simply replacing THF in the coordination sphere of the Lewis acid (main group element or transition metal) to give a new Lewis acid-base adduct, the Lewis acid and Lewis base act cooperatively on the THF molecule to effect ring-opening. These observations appear to suggest a relationship between systems that contravene Lewis' reactivity paradigm and steric congestion.

Further findings that contravened Lewis' axioms appeared in 1998 when Erker and coworkers[48] described the combination of the Lewis acid $B(C_6F_5)_3$ and the sterically encumbered ylide, Ph_3PCHPh. While this pairing initially led to the formation of the classical B–C Lewis adduct, $(Ph_3PCHPh)B(C_6F_5)_3$ at room temperature, this species was shown to undergo a new thermally-driven reaction in which the ylide attacks at the *para*-position of one of the fluorinated arene rings on boron, affording the zwitterionic salt $[Ph_3PCHPh(C_6F_4)BF(C_6F_5)_2]$ (Scheme 1.4).[48] This reaction was presumed to be a result of a thermally induced dissociation prompting a nucleophilic attack by the ylide on the arene with subsequent fluoride ion transfer to the boron center. Attack at the *para*-position was rationalized by the canonical resonance form of the free borane which generates a significant positive charge at the *para*-carbon atoms. Presumably, the formation of C–C and B–F bonds in the zwitterionic product is both irreversible and thermodynamically favorable.

In a related study, we examined the reactions of trityl borates with Lewis donors.[49] In most cases, classical Lewis acid–base adducts of the form $[(L)CPh_3][B(C_6F_5)_4]$ were formed. However, we noted that the reactions of the sterically encumbered phosphines,

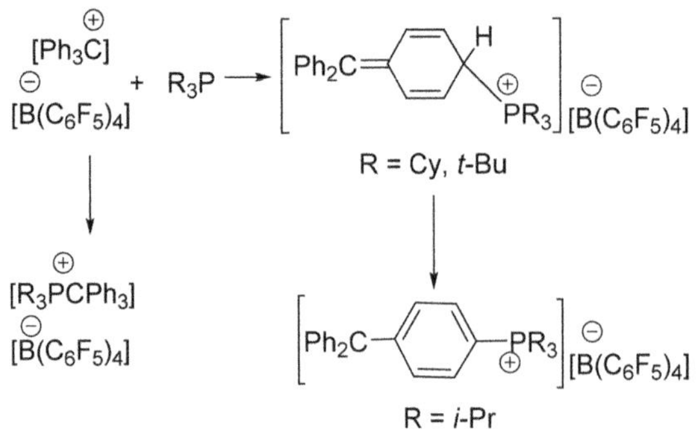

Scheme 1.5 Reactions of phosphines with the trityl cation.

PR_3, (R = i-Pr, Cy, t-Bu) did not form the expected adducts. Instead, nucleophilic aromatic substitution at a *para*-carbon of the carbocation occurs to give the species $[(R_3PC_6H_5) CPh_2][B(C_6F_5)_4]$ (R = Cy, t-Bu) or $[(i\text{-}Pr_3PC_6H_4)Ph_2CH][B(C_6F_5)_4]$ (Scheme 1.5).[49]

The above reports illustrate chemical outcomes that differ from the typical behavior of Lewis acid-base combinations. In these cases, the unique and unexpected reactivity was associated with systems in which the combinations of Lewis acids and bases are sterically demanding. While it is clear in retrospect that these observations foreshadowed important paradigm-altering findings, at the time of these reports, these findings seemed to be nothing more than idiosyncratic oddities.

1.9   Metal-free Activation of Dihydrogen

As we saw in Section 1.4, the ability of transition metals to active small molecules relies on the ability of the metal to act simultaneously as an electron acceptor and an electron donor. Indeed, it is because of these amphoteric characteristics that transition metals were deemed to be uniquely suited to catalysis. In Section 1.7 main group elements were described as Lewis acids and bases based on their capacity to accept and donate electrons. In Section 1.8, we saw that steric demands can prompt a deviation from the expected reactivity of classic Lewis acid-base combinations, allowing reactions with a third component. These observations prompted us to consider the following questions:

- Do the acceptor and donor properties need to reside on the same atom to achieve small molecule activation? Can a combination of a Lewis acceptor and donor affect small molecule activation?
- Can a transition metal be replaced with the combination of a main group electron acceptor and donor? Indeed, does one need a metal at all?

It appears that these notions were indirectly approached by the Piers group[50] in 2003 as they attempted to construct a main group species containing both protic and hydridic sites. In these efforts, they prepared the species $1,2\text{-}C_6H_4(NPh_2)(B(C_6F_5)_2)$ and delivered hydride to the N/B species generating the species $[Li(2\text{-crown-}4)][1,2\text{-}C_6H_4(NPh_2)(BH(C_6F_5)_2)]$.

Subsequent efforts to protonate the N-atom were, however, not successful. Nonetheless, subsequent reaction with the potent Brønsted acid known as Jutzi's acid $[(Et_2O)_2H]$ $[B(C_6F_5)_4]$ resulted in the liberation of dihydrogen and regenerated the amine-borane (Scheme 1.6).

In 2006, we[49] uncovered that zwitterionic salts of the form $[R_3P(C_6F_4)BF(C_6F_5)_2]$ could be readily prepared from the reactions of sterically demanding phosphines with $B(C_6F_5)_3$. These reactions proceed by *para*-attack of the borane by phosphine prompting P–C bond formation and fluoride transfer to boron. The isolation of these zwitterions was another example of reactivity contravening Lewis' axioms, as the anticipated classical Lewis acid–base adduct was not observed. We attributed this to steric hindrance disfavoring direct boron-phosphorus interactions, in a fashion closely related to the aforementioned system described by Erker[48] (Section 1.8).

Targeting an analogous product that could be further derivatized at phosphorus, the sterically encumbered secondary phosphine Mes_2PH was reacted with $B(C_6F_5)_3$ to give the zwitterion $[Mes_2PH(C_6F_4)BF(C_6F_5)_2]$. Subsequent reaction with $Me_2Si(H)Cl$, yielded $[Mes_2PH(C_6F_4)BH(C_6F_5)_2]$. This zwitterion was initially surprising to us, as it appeared to be a rare example of one that contains both a protic and a hydridic fragment. Indeed, heating of $[Mes_2PH(C_6F_4)BH(C_6F_5)_2]$ to 150 °C affected the liberation of dihydrogen affording the orange-red phosphine-borane $Mes_2P(C_6F_4)B(C_6F_5)_2$ (Scheme 1.7).[51] Spectroscopy of this latter phosphine-borane was consistent with a monomeric species, thus indicating the absence of an interaction between the Lewis basic phosphorus center with the Lewis acidic boron center in either a bimolecular or oligomeric fashion. The absence of such interactions was attributed to the steric demands of the substituents on both phosphorus and boron. Thus, $Mes_2P(C_6F_4)B(C_6F_5)_2$ could be considered

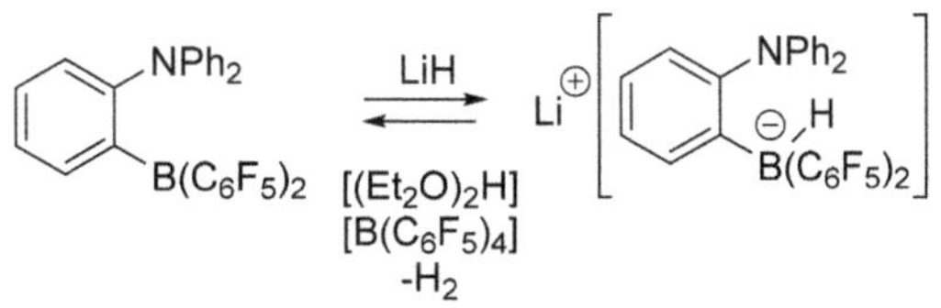

Scheme 1.6 Reaction of 1,2-$C_6H_4(NPh_2)(B(C_6F_5)_2)$.

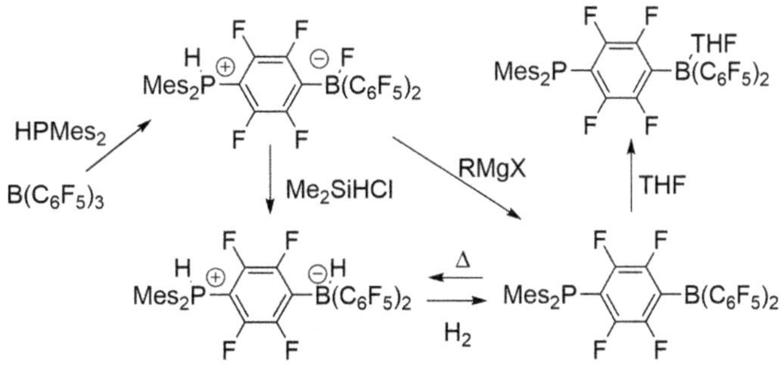

Scheme 1.7 Reactions of *para*-phosphine-borane species.

a combination of a Lewis acid and Lewis base in which the classical Lewis acid–base adduct formation is sterically inhibited. Nonetheless, the addition of an unencumbered donor like THF afforded the formation of a classical borane–THF Lewis acid–base adduct $Mes_2P(C_6F_4)B(C_6F_5)_2(THF)$ (Figure 1.5). The species $Mes_2P(C_6F_4)B(C_6F_5)_2$ was also accessible directly from $[Mes_2PH(C_6F_4)BF(C_6F_5)_2]$ *via* treatment with a Grignard reagent (Scheme 1.7).

The thermally induced loss of dihydrogen from the phosphonium-hydridoborate $[Mes_2PH(C_6F_4)BH(C_6F_5)_2]$ was not surprising given that this molecule contains both protic and hydridic groups. However, given that the loss of dihydrogen required heating and thus was perhaps thermodynamically uphill, it seemed reasonable to query if the reverse reaction could be facile. Following this logic, the phosphine-borane $Mes_2P(C_6F_4)B(C_6F_5)_2$ was exposed to 1 bar of dihydrogen at 25 °C. This resulted in the rapid formation of the zwitterionic salt (Scheme 1.7). The uptake of dihydrogen was visually obvious from the dramatic color change from the orange red (λ_{max}: 455 nm; $\varepsilon = 487$ L cm^{-1} mol^{-1}) of the phosphine-borane to the colorless solution of the zwitterionic phosphonium-borate salt (Figure 1.6). The color of the phosphine-borane presumably arises from weak π–donation from phosphorus to the electron-deficient boron center. However, this is disrupted upon quaternization of phosphorus and boron, and thus adducts or

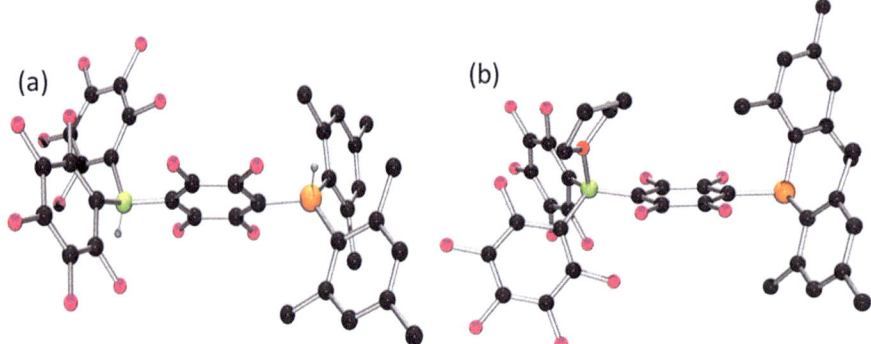

Figure 1.5   Crystal structures of (a) [Mes$_2$PH(C$_6$F$_4$)BH(C$_6$F$_5$)$_2$] and (b) Mes$_2$P(C$_6$F$_4$)B(C$_6$F$_5$)$_2$(THF).

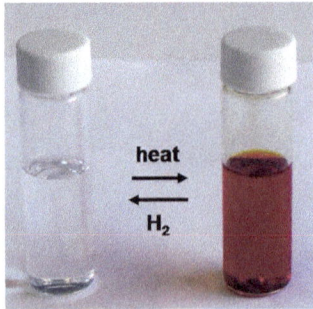

Figure 1.6   Solutions of [Mes$_2$PH(C$_6$F$_4$)BH(C$_6$F$_5$)$_2$] (left) and Mes$_2$P(C$_6$F$_4$)B(C$_6$F$_5$)$_2$ (right).

the phosphonium-borate zwitterion are colorless (Figure 1.6). The loss of dihydrogen from the zwitterion is unusual although we did note that the phosphine-borane adducts $R_2PH(BH_3)$ were known to thermally or catalytically eliminate dihydrogen to give cyclic and polymeric phosphinoboranes.[52,53] Nonetheless, the uptake of dihydrogen by the phosphine-borane, $Mes_2P(C_6F_4)B(C_6F_5)_2$, is a rare example of a non-transition metal system that reacts with dihydrogen.[54] Moreover, the interconversion of the phosphine-borane and the phosphonium-hydridoborate of this system makes it the first reported metal-free system capable of such reversible activation and release of dihydrogen.

Interestingly, increasing the basicity of the phosphorus center in the closely related species $t\text{-}Bu_2PH(C_6F_4)BH(C_6F_5)_2$ resulted in a thermally stable species. Even heating this derivative to 150 °C, failed to prompt dihydrogen release, inferring the lesser basicity of the phosphorus in the mesityl-substituted zwitterion enhances the Brønsted acidity of the corresponding phosphonium center thus facilitating reaction with the hydridic BH fragment and the liberation of dihydrogen.

1.10 Simple Phosphine-borane Combinations for the Activation of Dihydrogen

The above chemistry while precedent-setting, does involve a rather esoteric system. Nonetheless, the reactions of $[Mes_2PH(C_6F_4)BH(C_6F_5)_2]/Mes_2P(C_6F_4)B(C_6F_5)_2$ to release and take-up dihydrogen appeared to result from the inability of the phosphine Lewis base and the borane Lewis acid to form a classical Lewis acid-base adduct. This notion prompted another question:

- Could a similar heterolytic dihydrogen activation be achieved using simpler and perhaps more accessible combinations of Lewis acids and bases?

To this end, the combinations of the sterically encumbered phosphines R_3P (R = t-Bu, Mes) with $B(C_6F_5)_3$ were examined. These pairings showed no spectroscopic evidence of dative interactions, as the spectral data of the mixtures were indistinguishable from that of the individual constituents. Even upon cooling to -50 °C there was no spectroscopic evidence of the formation of the classical Lewis adducts.[55] Subsequent exposure to 1 bar of dihydrogen, resulted in the immediate precipitation of salts of the form $[R_3PH][HB(C_6F_5)_3]$ (R = Mes, t-Bu) (Scheme 1.8).[55] The formulations of these products

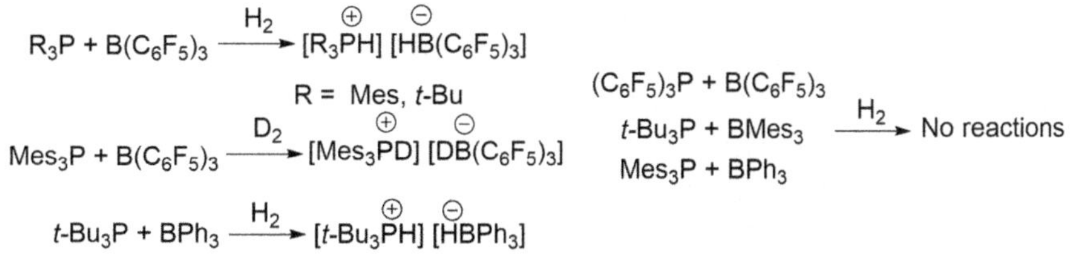

Scheme 1.8 Reaction of H_2 or D_2 with phosphine/borane combinations.

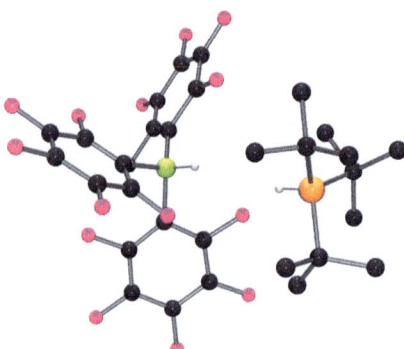

Figure 1.7 X-Ray crystal structure of [t-Bu$_3$PH][HB(C$_6$F$_5$)$_3$].

were confirmed by multi-nuclear NMR spectroscopy. In addition, the exposure of the combination of Mes$_3$P and B(C$_6$F$_5$)$_3$ to D$_2$ afforded [Mes$_3$PD][DB(C$_6$F$_5$)$_3$] (Scheme 1.8) as evidenced by the triplet in the ^{31}P NMR spectrum at −28.1 ppm with P–D coupling of 74 Hz, while the ^{2}D NMR spectrum showed the corresponding P–D resonance at 7.5 ppm and a broad B–D singlet at 3.8 ppm. X-Ray crystallographic data unambiguously confirmed the nature of the salt as [t-Bu$_3$PH][HB(C$_6$F$_5$)$_3$] (Figure 1.7) and further revealed the cations and anions are oriented in the solid-state such that the BH and HP fragments approach each other with a hydride-proton distance of 2.75 Å.

While heating [R$_3$PH][HB(C$_6$F$_5$)$_3$] to 150 °C did not liberate dihydrogen, the heterolytic cleavage of dihydrogen from the three-component reactions confirmed the ability of combinations of sterically encumbered Lewis acids and bases to affect the activation of dihydrogen. The impact of variations in the Lewis acidity and basicity of the components was further probed. The reaction of t-Bu$_3$P and BPh$_3$ with dihydrogen resulted in a slow reaction, affording [t-Bu$_3$PH][HBPh$_3$] in 33% yield (Scheme 1.8). Similarly, the reactions of Mes$_3$P and BPh$_3$, (C$_6$F$_5$)$_3$P and B(C$_6$F$_5$)$_3$ or t-Bu$_3$P and BMes$_3$ resulted in no observable dihydrogen activation. These results inferred that despite apparent steric frustration, an aggregate Lewis acidity and basicity appears required to affect the heterolytic cleavage of dihydrogen.

1.11 Intramolecular Phosphine-borane Species

Following the above discoveries, efforts to uncover new intramolecular linked phosphine-borane systems capable of the activation of dihydrogen were explored.[56] The Erker group was motivated by a Tilley group report that phosphine-boranes of the form [Ph$_2$PCH$_2$CH$_2$BR$_2$]$_n$ (R^1 = Cy, 9-BBN) could be derived from regioselective hydroboration,[57] and used as 'ambiphilic' ligands.[58] Targeting the inclusion of a more acidic borane in an ethylene-linked P/B species, Erker and coworkers treated Ph$_2$PCH=CH$_2$ with 'Piers' borane' HB(C$_6$F$_5$)$_2$.[59–63] However, this gave the classical Lewis acid–base adduct, CH$_2$=CHPPh$_2$(HB(C$_6$F$_5$)$_2$) (Scheme 1.9).[64] In contrast, the corresponding reactions of allyl- and butenyl-phosphines, CH$_2$=CHCH$_2$PR$_2$ (R = Ph, t-Bu) and CH$_2$=CHCH$_2$CH$_2$PPh$_2$,

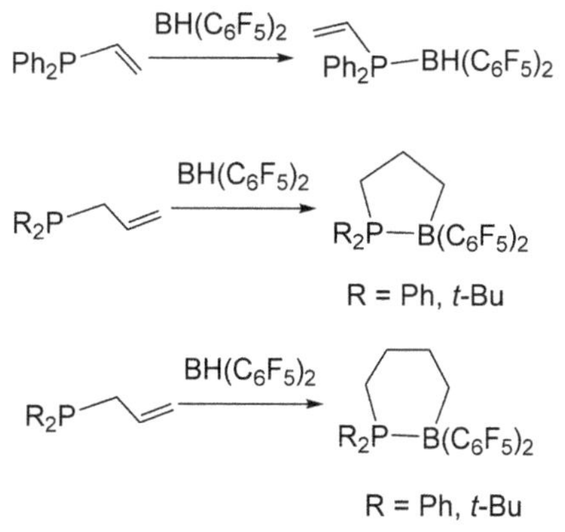

Scheme 1.9 Reactions of Piers' borane and vinyl phosphines.

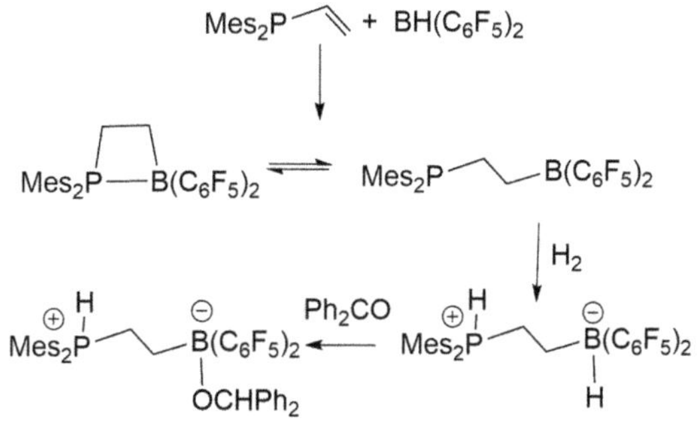

Scheme 1.10 Synthesis and reaction of the zwitterionic salt [Mes$_2$PHCH$_2$CH$_2$BH(C$_6$F$_5$)$_2$].

provided the desired hydroboration with HB(C$_6$F$_5$)$_2$ giving the phosphine-borane products, R$_2$P(CH$_2$)$_3$B(C$_6$F$_5$)$_2$ (R = Ph, t-Bu) and R$_2$P(CH$_2$)$_4$B(C$_6$F$_5$)$_2$ (R = Ph, t-Bu), respectively (Scheme 1.9). These species formed strong intramolecular P/B Lewis acid–base adducts resulting from the formation of the robust and unreactive five and six-membered rings, respectively.

Using a similar synthetic protocol and a more sterically demanding phosphine precursor, the species Mes$_2$PCH=CH$_2$ was treated with HB(C$_6$F$_5$)$_2$ to produce Mes$_2$PCH$_2$CH$_2$B(C$_6$F$_5$)$_2$ (Scheme 1.9).[56] Spectroscopic and computational data reveal that Mes$_2$PCH$_2$CH$_2$B(C$_6$F$_5$)$_2$ exists in an equilibrium between its 'closed' and 'open' form. In its minimum energy configuration, this molecule features a four-membered heterocyclic structure with a weak P⋯B interaction (2.21 Å) and π–π-stacking interactions between the electron-poor C$_6$F$_5$ rings and the electron-rich mesityl rings. In the open form, calculations identified gauche and antiperiplanar conformational minima that are 8 to 12 kcal mol^{-1} higher in energy than the closed conformation, respectively.[65] Exposure of this species to 1.5 bar of dihydrogen at 25 °C resulted in the rapid precipitation of the zwitterionic salt [Mes$_2$PHCH$_2$CH$_2$BH(C$_6$F$_5$)$_2$] (Scheme 1.10). This heterolytic cleavage of dihydrogen was confirmed spectroscopically and the corresponding reaction with D_2 gave the corresponding deuterated zwitterion, $[Mes_2PDCH_2CH_2BD(C_6F_5)_2]$. The zwitterionic salt $[Mes_2PHCH_2CH_2BH(C_6F_5)_2]$ was shown to subsequently react with benzophenone stoichiometrically to give the alkoxy-borate zwitterion $[Mes_2PHCH_2 CH_2B(OCH_2Ph)(C_6F_5)_2]$ (Scheme 1.10).[66]

These results further affirm that both inter- and intramolecular B/P systems with the appropriate steric demands can generate FLPs capable of the activation of dihydrogen. With careful tuning of the cumulative basicity/acidity, such activation of dihydrogen can be reversible. Given these demonstrations, this prompted us to probe the generality of the reactivity of FLPs with other small molecules.

1.12 Reactions of 'Frustrated Lewis Pairs (FLPs)' with Olefins

Our initial thought was to test the reactivity of FLPs with olefins. To that end, the combination of t-Bu$_3$P and B(C$_6$F$_5$)$_3$ was treated with ethylene (1 bar) at 25 °C. This prompted the precipitation of a product that was formulated by NMR spectroscopy as $[t$-Bu$_3$P(C$_2$H$_4$)B(C$_6$F$_5$)$_3]$. This was further confirmed by crystallographic studies (Scheme 1.11, Figure 1.8).[67] The data affirmed the formation of a zwitterion in which the phosphonium and borate fragments are linked by the two carbon atoms derived from ethylene. Similar products, $[t$-Bu$_3$PCH(R)CH$_2$B(C$_6$F$_5$)$_3]$ (R = Me, C$_4$H$_9$) were derived from the corresponding reactions of propene or 1-hexene. In these cases, additions were regioselective with the P-atom adding to the secondary olefinic carbon (Scheme 1.11).

Such additions of phosphine/borane combinations to olefins were also shown to proceed in an intramolecular fashion. To this end, the species $CH_2=CH(CH_2)_3PR_2$ (R = t-Bu, Mes) were combined with B(C$_6$F$_5$)$_3$ prompting an intramolecular cyclization and the formation of phosphonium borates $[R_2PCH(C_3H_6)CH_2B(C_6F_5)_3]$ (R = t-Bu, Mes) (Scheme 1.11, Figure 1.8).[67] Again, crystallographic data confirmed the regioselective addition of the phosphorus atom to the secondary olefinic carbon.

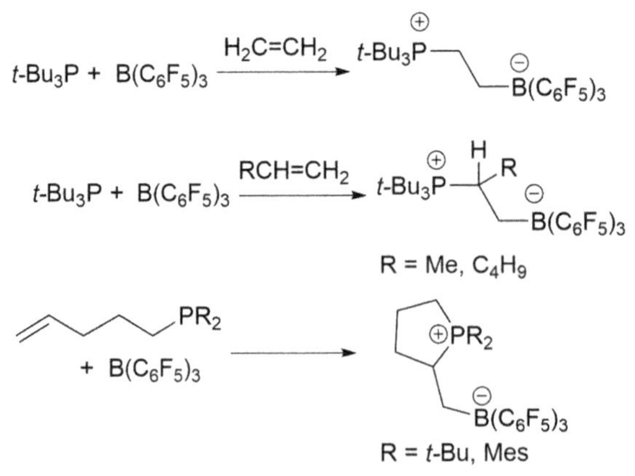

Scheme 1.11 Reactions of phosphine/borane combinations with olefins.

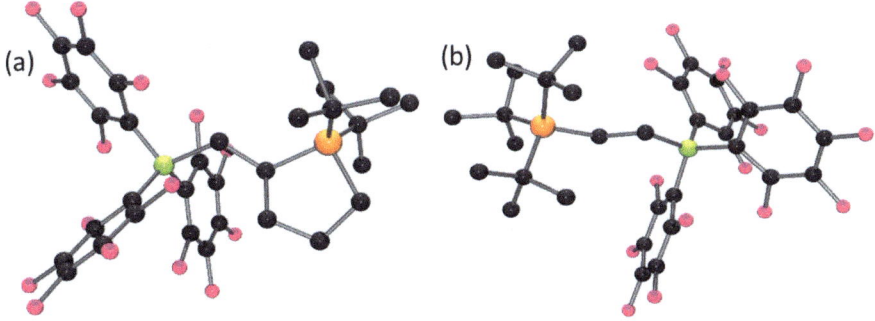

Figure 1.8 Crystal structures of (a) [t-Bu$_2$PCH(C$_3$H$_6$)CH$_2$B(C$_6$F$_5$)$_3$] and (b) [t-Bu$_3$P(C$_2$H$_4$)B(C$_6$F$_5$)$_3$].

Mechanistically these reactions were proposed to proceed *via* Lewis acidic activation of the olefin, prompting nucleophilic attack by the phosphine at the more-substituted carbon atom. Such Lewis acid–olefin interactions had been previous computed for ethylene–alane[68] and ethylene–borane adducts.[68,69] Interestingly, in these reactions, combinations of any two of the Lewis acid, Lewis base, and olefin led to no reaction, whereas the combination of all three led to the formation of the zwitterionic addition products. It was this work that illustrated the broader potential for the activation of small molecules by combinations of sterically encumbered Lewis acids and bases. As a result, in reporting this chemistry, we coined these systems as '*frustrated Lewis pairs (FLPs)*' to describe these uniquely reactive systems.

1.13   The End of the Beginning

The above chemistry establishes that mixtures of sterically frustrated Lewis acid/base combinations, or FLPs, could react both with hydrogen and olefins. In some fashion, this mimics the reactivity of transition metal centers, although the donor and acceptor properties normally residing on a single metal center are now separated into the Lewis acid and base components of the FLP. Of course, for such combinations to be reactive, conventional Lewis acid–base adduct formation must be sterically inhibited.

It is also important to recognize that the articulation of the ability of FLPs to activate dihydrogen was surprising. It contravened the chemical dogma that had been held for over 100 years regarding the presumed unique ability of transition metals to activate dihydrogen. Moreover, this finding immediately prompted a range of new questions regarding the mechanism and the potential of FLPs in catalytic metal-free hydrogenations. Questions were also raised about the generality of the concept and the ability of FLPs to activate a broad range of other small molecules. Over the last decade and a half, many researchers have probed these and related ideas. In addition, other researchers have exploited FLPs in highly creative ways, applying FLP chemistry to areas that were not even conceived of in the early years.[70–73] Indeed, the broadening exploitation of the concept of FLPs has been extraordinary.[74]

This text focuses on the developments derived from the initial finding of FLP-dihydrogen activation. This one aspect of FLP chemistry has garnered terrific interest and has developed explosively, to afford a myriad of metal-free systems capable of activating dihydrogen and strategies towards the metal-free reduction of many substrates. This has further evolved to provide elegant approaches to metal-free enantioselective hydrogenations as well as even broadener applications in inorganic, green, bioinorganic, and catalytic chemistry. We provide this text in the hope that it will serve two purposes. Firstly, it is intended to provide an introduction and understanding of the origin, fundamentals, and scope of this new reactivity paradigm to those unfamiliar with the area. Secondly, we hope it will encourage broader, creative applications of metal-free hydrogenation catalysis.

References

1. P. Sabatier, *Compt. Rend.*, 1897, **124**, 616–618.
2. P. Sabatier, *Ind. Eng. Chem.*, 1926, **18**, 1005–1008.
3. F. Haber, *Naturwissenschaften*, 1922, 1041–1049.
4. G. Wilke, *Angew. Chem.*, 2003, **42**, 5000–5008.
5. G. Natta, *J. Polym. Sci.*, 1955, **16**, 143–154.
6. G. Natta, *Makromol. Chem.*, 1955, **16**, 213–237.
7. G. Natta, P. Pino, P. Corradini, F. Danusso, E. Mantica, G. Mazzanti and G. Moraglio, *J. Am. Chem. Soc.*, 1955, **77**, 1708–1710.
8. G. Natta, P. Pino and G. Mazzanti, *Angew. Chem., Int. Ed.*, 1955, **67**, 759.
9. J. A. Osborn, F. H. Jardine, J. F. Young and G. Wilkinson, *J. Chem. Soc. A*, 1966, 1711–1732.
10. P. S. Hallman, D. Evans, J. A. Osborn and G. Wilkinson, *Chem. Commun.*, 1967, 305–306.
11. W. S. Knowles, *Angew. Chem., Int. Ed.*, 2002, **41**, 1998–2007.
12. R. Noyori, *Angew. Chem., Int. Ed.*, 2002, **41**, 2008–2022.
13. K. B. Sharpless, *Angew. Chem., Int. Ed.*, 2002, **41**, 2024–2032.
14. R. H. Grubbs, *Angew. Chem., Int. Ed.*, 2006, **45**, 3760–3765.
15. R. Schrock, *Angew. Chem., Int. Ed.*, 2006, **45**, 3748–3759.
16. Y. Chauvin, *Angew. Chem., Int. Ed.*, 2006, **45**, 3740–3747.
17. E. I. Negishi, *Angew. Chem., Int. Ed.*, 2011, **50**, 6738–6764.
18. A. Suzuki, *Angew. Chem., Int. Ed.*, 2011, **50**, 6722–6737.
19. R. F. Heck, *J. Am. Chem. Soc.*, 1968, **90**, 5518–5526.
20. J. Védrine, *Catalysts*, 2017, 7, 341.
21. R. H. Morris, *Coord. Chem. Rev.*, 2008, **252**, 2381–2394.
22. G. J. Kubas, R. R. Ryan, B. I. Swanson, P. J. Vergamini and H. J. Wasserman, *J. Am. Chem. Soc.*, 1984, **106**, 451–452.
23. M. J. S. Dewar, *Bull. Soc. Chim. Fr.*, 1951, **18**, C71.
24. J. Chatt and L. A. Duncanson, *J. Chem. Soc.*, 1953, 2939–2947.
25. M. R. Elsby and R. T. Baker, *Chem. Soc. Rev.*, 2020, **49**, 8933–8987.
26. S. Zaib and I. Khan, *Curr. Org. Chem.*, 2020, **24**, 1775–1792.
27. G. N. Lewis, *Valence and the Structure of Atoms and Molecules*, Chemical Catalogue Company, Inc., New York, 1923.
28. J. N. Brønsted, *Recl. Trav. Chim. Pays-Bas*, 1923, **42**, 718–728.
29. F. A. Cotton, G. Wilkinson and C. A. Murillo, *Advanced Inorganic Chemistry*, John Wiley & Sons Canada, Ltd., Toronto, 6th edn, 1999.
30. *Lewis Acids in Organic Synthesis*, ed. H. Yamamoto, Wiley-VCH, Weinheim, 2000.
31. H. C. Brown, H. I. Schlesinger and S. Z. Cardon, *J. Am. Chem. Soc.*, 1942, **64**, 325–329.
32. H. C. Brown and B. Kanner, *J. Am. Chem. Soc.*, 1966, **88**, 986–992.
33. G. Wittig and A. Ruckert, *Liebigs Ann. Chem.*, 1950, **566**, 101–113.
34. G. Wittig and E. Benz, *Chem. Ber.*, 1959, **92**, 1999–2013.
35. W. Tochtermann, *Angew. Chem., Int. Ed.*, 1966, **5**, 351–371.

36. T. L. Breen and D. W. Stephan, *Inorg. Chem.*, 1992, **31**, 4019–4022.
37. L. R. Avens, D. M. Barnhart, C. J. Burns and S. D. McKee, *Inorg. Chem.*, 1996, **35**, 537–539.
38. M. P. C. Campello, A. Domingos and I. Santos, *J. Organomet. Chem.*, 1994, **484**, 37–46.
39. W. J. Evans, J. T. Leman, J. W. Ziller and S. I. Khan, *Inorg. Chem.*, 1996, **35**, 4283–4291.
40. A. Mommertz, R. Leo, W. Massa, K. Harms and K. Dehnicke, *Z. Anorg. Allg. Chem.*, 1998, **624**, 1647–1652.
41. Z. Y. Guo, P. K. Bradley and R. F. Jordan, *Organometallics*, 1992, **11**, 2690–2693.
42. M. Polamo, I. Mutikainen and M. Leskelä, *Acta Crystallogr., Sect. C*, 1997, **C53**, 1036–1037.
43. M. Gomez-Saso, D. F. Mullica, E. Sappenfield and F. G. A. Stone, *Polyhedron*, 1996, **15**, 793–801.
44. J. P. Campbell and W. L. Gladfelter, *Inorg. Chem.*, 1997, **36**, 4094–4098.
45. T. Chivers and G. Schatte, *Eur. J. Inorg. Chem.*, 2003, 3314–3317.
46. S. M. Kunnari, R. Oilunkaniemi, R. S. Laitinen and M. Ahlgren, *Dalton Trans.*, 2001, 3417–3418.
47. G. C. Welch, J. D. Masuda and D. W. Stephan, *Inorg. Chem.*, 2006, **45**, 478–480.
48. S. Doering, G. Erker, R. Fröhlich, O. Meyer and K. Bergander, *Organometallics*, 1998, **17**, 2183–2187.
49. L. Cabrera, G. C. Welch, J. D. Masuda, P. R. Wei and D. W. Stephan, *Inorg. Chim. Acta*, 2006, **359**, 3066–3071.
50. R. Roesler, W. E. Piers and M. Parvez, *J. Organomet. Chem.*, 2003, **680**, 218–222.
51. G. C. Welch, R. R. S. Juan, J. D. Masuda and D. W. Stephan, *Science*, 2006, **314**, 1124–1126.
52. T. L. Clark, J. M. Rodezno, S. B. Clendenning, S. Aouba, P. M. Brodersen, A. J. Lough, H. E. Ruda and I. Manners, *Chem. - Eur. J.*, 2005, **11**, 4526–4534.
53. C. A. Jaska and I. Manners, *J. Am. Chem. Soc.*, 2004, **126**, 9776–9785.
54. G. H. Spikes, J. C. Fettinger and P. P. Power, *J. Am. Chem. Soc.*, 2005, **127**, 12232–12233.
55. G. C. Welch and D. W. Stephan, *J. Am. Chem. Soc.*, 2007, **129**, 1880–1881.
56. P. Spies, G. Erker, G. Kehr, K. Bergander, R. Fröhlich, S. Grimme and D. W. Stephan, *Chem. Commun.*, 2007, 5072–5074.
57. A. Fischbach, P. R. Bazinet, R. Waterman and T. D. Tilley, *Organometallics*, 2008, **27**, 1135–1139.
58. F.-G. Fontaine, J. Boudreau and M.-H. Thibault, *Eur. J. Inorg. Chem.*, 2008, 5439–5454.
59. D. J. Parks, W. E. Piers, M. Parvez, R. Atencio and M. J. Zaworotko, *Organometallics*, 1998, **17**, 1369–1377.
60. D. J. Parks, R. E. v. H. Spence and W. E. Piers, *Angew. Chem., Int. Ed.*, 1995, **34**, 809–811.
61. W. E. Piers and T. Chivers, *Chem. Soc. Rev.*, 1997, **26**, 345–354.
62. R. E. V. Spence, D. J. Parks, W. E. Piers, M. A. Macdonald, M. J. Zaworotko and S. J. Rettig, *Angew. Chem., Int. Ed.*, 1995, **34**, 1230–1233.
63. R. E. V. Spence, W. E. Piers, Y. M. Sun, M. Parvez, L. R. MacGillivray and M. J. Zaworotko, *Organometallics*, 1998, **17**, 2459–2469.
64. P. Spies, G. Kehr, K. Bergander, B. Wibbeling, R. Fröhlich and G. Erker, *Dalton Trans.*, 2009, 1534–1541.
65. C. M. Mömming, S. Fromel, G. Kehr, R. Fröhlich, S. Grimme and G. Erker, *J. Am. Chem. Soc.*, 2009, **131**, 12280–12289.
66. P. Spies, G. Kehr, S. Kehr, R. Fröhlich and G. Erker, *Organometallics*, 2007, **26**, 5612–5620.
67. J. S. J. McCahill, G. C. Welch and D. W. Stephan, *Angew. Chem., Int. Ed.*, 2007, **46**, 4968–4971.
68. P. Tarakeshwar and K. S. Kim, *J. Phys. Chem. A*, 1999, **103**, 9116–9124.
69. P. Tarakeshwar, S. J. Lee, J. Y. Lee and K. S. Kim, *J. Phys. Chem. B*, 1999, **103**, 184–191.
70. D. W. Stephan, *Science*, 2016, **354**, aaf7229.
71. D. W. Stephan and G. Erker, *Angew. Chem., Int. Ed.*, 2015, **54**, 6400–6441.
72. D. W. Stephan, *J. Am. Chem. Soc.*, 2015, **137**, 10018–10032.
73. D. W. Stephan, *Acc. Chem. Res.*, 2015, **48**, 306–316.
74. A. R. Jupp and D. W. Stephan, *Trends Chem.*, 2019, **1**, 35–48.

2 The Nature of Frustrated Lewis Pairs

2.1 Chapter Overview

Having described the seminal papers in the chemistry of phosphine/borane FLPs in the initial chapter, we now consider the features and nature of these combinations of Lewis acids and bases. Initially, we discuss the mechanism in terms of literature precedent and then in terms of subsequent computational studies. Experimental work probing the kinetics and thermodynamics are described, while the reactivity of select FLPs *via* a radical pathway is also considered. Further insights about the nature of FLPs have been obtained from studies varying the acidic and basic components. Several systems illustrate that the notion of FLP reactivity is not limited to systems where steric encumbrance precludes adduct formation. It is critical to understand this broader perspective, as this will impact the design of FLP catalysts going forward.

2.2 Initial Thoughts on the Mechanism

Perhaps the first, most pressing question prompted by the observation of dihydrogen activation by FLPs related to the mechanism of action. The combination of a Lewis acid, Lewis base, and dihydrogen seemed to suggest a three-component reaction. This was further supported by the absence of reaction between any combination of two of these reagents. In efforts to provide some rationalization for this initially perplexing observation, early rationales were based on the analogy to transition metal-dihydrogen interactions. Thus, the notion of activation of dihydrogen by a Lewis acid or base was considered. Previous computational studies supported the existence of the weak borane–dihydrogen adduct, $(\eta^2\text{-H}_2)BH_3$.[1–4] Interestingly, Moroz and Sweany speculated that an analogous dihydrogen adduct of the photochemically generated radical, $\cdot BBr_2$ lead to the formation of $HBBr_2$.[5] A similar initial mechanism involving the interaction of dihydrogen with the Lewis acid, $B(C_6F_5)_3$, was proposed to affect the polarization of dihydrogen facilitating protonation of the Lewis basic phosphine (Figure 2.1).

A Primer in Frustrated Lewis Pair Hydrogenation: Concepts to Applications
By Douglas W. Stephan
© Douglas W. Stephan 2022
Published by the Royal Society of Chemistry, www.rsc.org

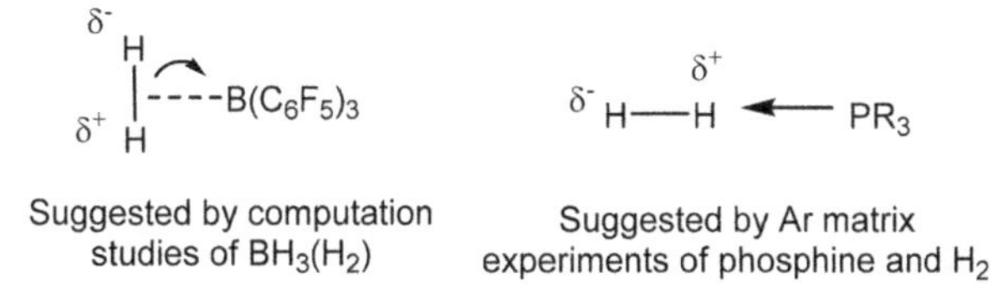

Suggested by computation   Suggested by Ar matrix
studies of BH₃(H₂)   experiments of phosphine and H₂

Figure 2.1 Interactions of borane and phosphine with dihydrogen based on previous literature.

Despite the apparent logic and precedence of this proposition, attempts to observe the borane–dihydrogen adduct, $(H_2)B(C_6F_5)_3$, by NMR methods were unsuccessful.

An alternative possibility focused on the initial interaction of phosphine with dihydrogen. It is noteworthy in this regard, that Sweany and coworkers[6] described experimental data derived from the combination of phosphines and dihydrogen in an argon matrix at very low temperatures that inferred the generation of van der Waal complexes. These interactions presumably affect the polarization of dihydrogen *via* phosphine donation to the σ*-orbital of the dihydrogen molecule (Figure 2.1). Were such interactions present in the reactions of FLPs with dihydrogen, this could prompt hydride delivery to a Lewis acid, thus affecting the observed heterolytic cleavage of dihydrogen.

2.3 Computational Studies

A clearer understanding of the mechanism of dihydrogen activation was provided by computational studies initially performed by Papai and coworkers.[7–10] Perhaps contrary to initial suppositions, the notion of a van der Waals complex[15] of dihydrogen with $B(C_6F_5)_3$ is unlikely to contribute to the dihydrogen activation pathway. Rather, these studies proposed the initial formation of an 'encounter complex' resulting from the close approach of the Lewis acid and base in which steric conflicts preclude the formation of a classical Lewis acid–base adduct. In the case of the FLP, t-Bu₃P/B$(C_6F_5)_3$, computations show a shallow thermodynamic well in which the encounter complex resides with an association energy of −11.5 kcal mol^{-1}. This is only approximately half the value typically observed for the formation of a classical Lewis adduct.[11] Dispersion forces resulting from the multiple interactions of C–H bonds in the t-butyl groups and the fluorine atoms in the borane stabilize the encounter complex with a phosphorus–boron distance of about 4.2 Å. The encounter complex residing in such a shallow thermodynamic well suggests that it is short-lived. However, access to such a species provides a rational bimolecular reaction pathway involving the encounter complex and dihydrogen. Thus, these computational findings resolve the perplexing situation in which the FLP activation of dihydrogen appeared (if only superficially) to be a trimolecular process. Papai also computed the geometry of the transition state which involved a linear B–H–H–P fragment (Figure 2.2). The transition state was computed to be only 10.4 kcal mol^{-1} above the encounter complex, as the transition state maintains the C–H–F dispersion interactions. In the transition state, the dihydrogen molecule is polarized and subsequently undergoes heterolytic cleavage to afford the observed phosphonium hydridoborate product in an exothermic process ($\Delta E = -26.3$ kcal mol^{-1}).

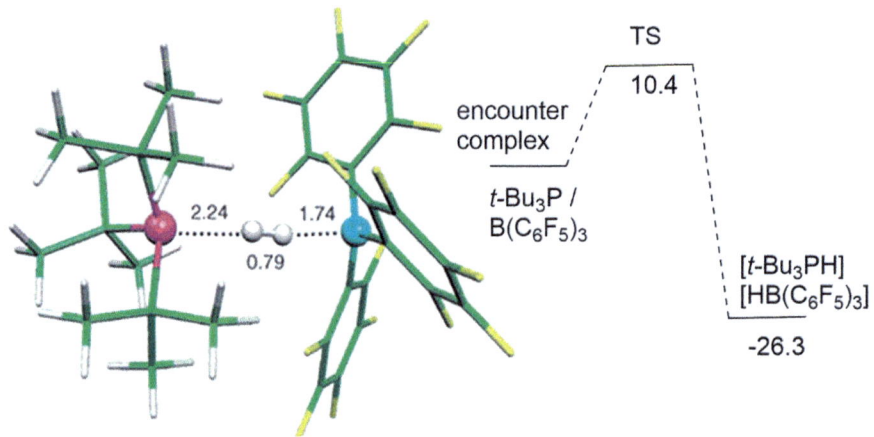

Figure 2.2   Depiction of t-Bu$_3$P/B(C$_6$F$_5$)$_3$ 'encounter complex' with dihydrogen computed by Papai.[7] Reproduced from ref. 7 with permission from John Wiley and Sons, Copyright © 2008 WILEY-VCH Verlag GmbH & Co. KGaA, Weinheim.

Papai's group[12] also probed the details of the activation of dihydrogen. They proposed that it proceeds by an electron-transfer mechanism in which the encounter complex and dihydrogen approach while the frontier orbitals are unchanged. Furthermore, they suggest that the activation energy leads to the perturbation of these orbitals, resulting in a concerted electron transfer process *via* donation of the Lewis base lone pair to the σ* orbital of dihydrogen and from the σ orbital of dihydrogen to the acceptor orbital of the Lewis acid. In addition to this electron transfer, the activation energy must also account for the structural distortions at the boron and phosphorus centers.[9]

In a further study, Grimme *et al.*,[13] used a higher level of theory and accounted for dispersion effects in considering the activation of dihydrogen by the intramolecular FLP Mes$_2$PCH$_2$CH$_2$B(C$_6$F$_5$)$_2$, as well as the intermolecular FLP system derived from t-Bu$_3$P/B(C$_6$F$_5$)$_3$. Like Papai's computations, the computed transition states exhibit C–H–F interactions and a slightly elongated H–H bond in the dihydrogen fragment of 0.74 to 0.79 Å consistent with an asymmetric and concerted activation of dihydrogen by the FLP. However, for the interaction of dihydrogen with t-Bu$_3$P/B(C$_6$F$_5$)$_3$ or Mes$_2$PCH$_2$CH$_2$B(C$_6$F$_5$)$_2$, the dihydrogen molecule is oriented off the phosphorus–boron axis and, contrary to the Papai findings, this orientation suggests the H–H σ-bond interacts with the vacant orbital on boron while the lone pair of phosphorus donate to the dihydrogen σ*-orbital (Figure 2.3). Once in this position, the reaction proceeds *without a barrier* as this orientation polarizes dihydrogen resulting in the partial cleavage of dihydrogen and increases electron density in the dihydrogen σ orbital proximal to the Lewis acid. Regardless of the precise geometric details, the interactions leading to the activation of dihydrogen by an FLP are reminiscent of the interactions of a transition metal center with dihydrogen where the metal acts both as σ-acceptor and π-back-bonding donor. Of course, in contrast to metal-based activation, the acceptor and donor orbitals in FLP systems reside on separate molecules.

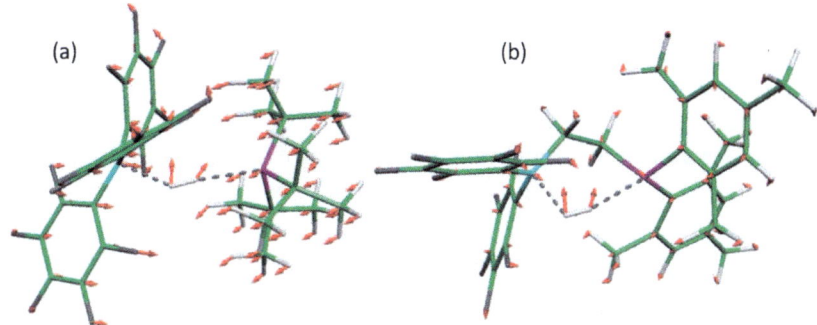

Figure 2.3 Computed interactions of dihydrogen with (a) t-Bu$_3$P/B(C$_6$F$_5$)$_3$ and (b) Mes$_2$PCH$_2$ CH$_2$B(C$_6$F$_5$)$_2$ computed by Grimme et al.[13] Reproduced from ref. 13 with permission from John Wiley and Sons, Copyright © 2010 WILEY-VCH Verlag GmbH & Co. KGaA, Weinheim.

Grimme and Erker[13] noted the approach of the phosphorus and boron in the encounter complex generated a strongly polar environment and showed computationally that the FLP can be replaced by an electric field. Further, they showed a field strength of about 0.05–0.06 atomic units, providing sufficient potential to cleave dihydrogen. Interestingly, larger field strengths (>0.06 a.u.) were predicted to alter the mechanism from a biradical-like to a zwitterionic process.[14] This notion was related to the experimental observation that dihydrogen could be cleaved using a weaker Lewis acid if one employed a stronger base, as these two features cumulatively generate a sufficiently strong electric field. Moreover, Grimme and Erker also pointed out the noncovalent interactions between the bulky substituents in the acid and base are flexible, providing ready entrance of dihydrogen to the interior of the encounter complex derived from the FLP.

The groups of Privalov[15] and Papai[16] also exploited molecular dynamics simulations to study the FLP, t-Bu$_3$P/B(C$_6$F$_5$)$_3$. They showed the onset of dihydrogen activation begins at a much larger boron···phosphorus distance than previously thought while inferring that about 2% of the total amount of phosphine and borane are associated and, of those, only about 0.5% are properly oriented for dihydrogen activation. Similarly, studies of molecular dynamics for intramolecular FLPs have also recently appeared.[17] Further computational work[18] probed the time scales of nuclear motion and the time-dependence of the donor–acceptor interactions in the reaction of the encounter complex with dihydrogen, and this was supported by femtosecond spectroscopy.[19]

2.4   Pertinent Precedent

It is important to emphasize the direct analogy of the above computed mechanistic models to that described by Piers some years prior. In 1996, Piers and coworkers[20] reported the catalytic mediation of the hydrosilylation of ketones, aldehydes, and esters by B(C$_6$F$_5$)$_3$. These authors showed the reactions were first order in silane. While the k_{obs} should be expected to show first-order or zero dependence on substrate concentration, the experimental data showed a decrease in k_{obs} as ketone concentration rises. This observation led Piers to conclude that the role of the Lewis acid was not to activate the carbonyl substrate but rather that $B(C_6F_5)_3$ activates the Si–H of the silane prompting nucleophilic attack of silicon by the carbonyl species (Figure 2.4). Piers pointed out that this mechanism led to a paradox in that the least basic substrates are reduced more rapidly, as carbonyl binding to borane, while non-productive, competes with Si–H activation.

The mechanistic finding by Piers was surprising at the time and indeed the broader implications and truly seminal nature of this study were not appreciated for over a decade. The Piers mechanism was further substantiated in a series of elegant experiments some years later by Oestreich and coworkers (Scheme 2.1).[21] These researchers utilized a chiral silane in a borane-mediated hydrosilylation. Characterization of the product affirmed inversion of the chirality at silicon consistent with the backside attack of the silane in an S_N2-type fashion, which is activated by $B(C_6F_5)_3$.

In the context of FLPs, the mechanism proposed by Piers is directly analogous to that proposed above for dihydrogen activation by FLPs. Indeed, one can view the Si–H bond activation as arising from the cooperative action of the Lewis acidic borane and the basic carbonyl on the Si–H bond. A final piece of proof was further offered by Piers *et al.* in 2014[22,23] when they reported the exceptionally strong Lewis acid 1,2,3-tris(pentafluorophenyl)-4,5,6,7-tetrafluoro-1-boraindene was able to form an isolable adduct with triethylsilane (Scheme 2.2). While this species exists in equilibrium with free silane and borane, low-temperature isolation afforded full spectroscopic and crystallographic characterization revealing a B···H distance of 1.46(2) Å. Subsequent study of the reactivity with nucleophiles demonstrated the role of such Lewis acid activation of Si–H

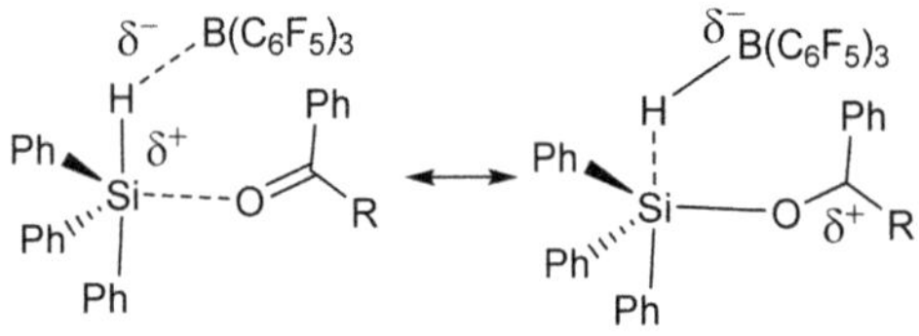

Figure 2.4 Nature of intermediate proposed by Piers.

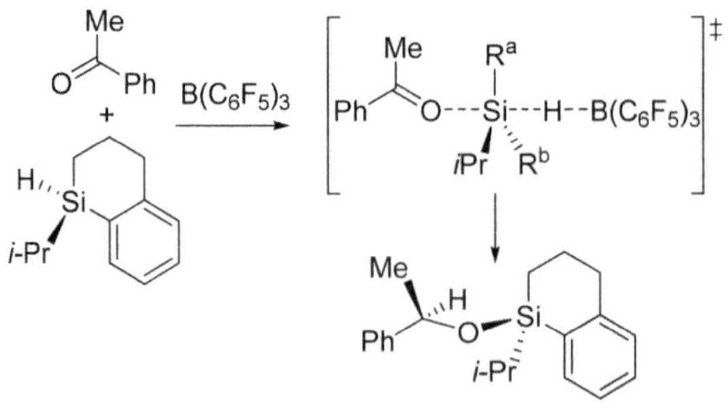

Scheme 2.1 The Oestreich experiment affirming the Piers mechanism for hydroboration of ketones.

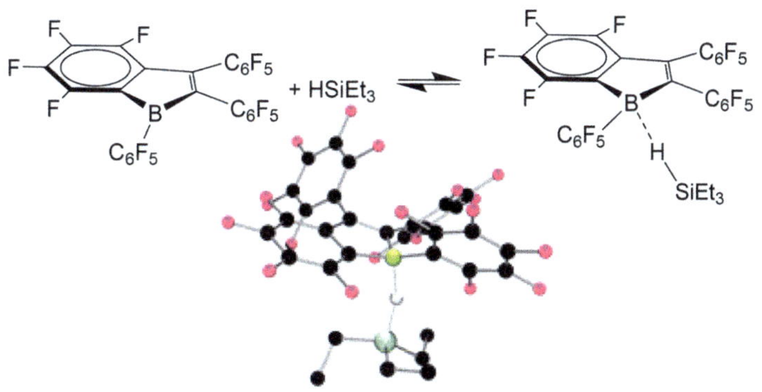

Scheme 2.2 Equilibrium governing the silane adduct of Piers' boraindene. POV-ray depiction of the adduct, C: black, F: pink, B: yellow-green, H: gray, Si: light green. Hydrogen atoms except for the Si–H have been omitted for clarity.

bonds in FLP hydrosilylation. Interestingly, the next year, Chen and Chen[24] reported the Al-based Lewis acid, $Al(C_6F_5)_3$ formed the analogous silane adduct $(Et_3SiH)Al(C_6F_5)_3$ with an Al$\cdots$H distance of 1.865(16) Å.

2.5 Experimental Evidence for Encounter Complexes

Statistical and entropic reasons disfavor the possibility of a termolecular transition state. Indeed, the computational studies described above infer that an 'encounter complex' formed by the Lewis acid and base reacts with dihydrogen in a bimolecular fashion. Such encounter complexes are short-lived, as they are stabilized only by weak dispersion forces in addition to the inherent electrostatic attraction. This view is also consistent with the enhanced reactivity of intramolecular FLP systems that are pre-organized for interaction with dihydrogen.

In general, the short lifetime and low concentration preclude the experimental observation of encounter complexes. However, Macchioni *et al.*[25] studied the FLP, t-Bu$_3$P/ $B(C_6F_5)_3$ by ^{1}H/^{19}F, with 2D HOESY experiments demonstrating the correlation between the fluorine atoms and the hydrogen atoms in the t-Bu substituents. Additional diffusion NMR experiments inferred an associative process that is slightly endergonic with a $\Delta G^0(298\ \text{K}) = +0.4\ \text{kcal mol}^{-1}$. These observations are consistent with the short lifetimes and low concentrations of encounter complexes.[25]

Some additional support for the notion of an encounter complex was provided by a neutron study[26] of the product of the reaction of the FLP derived from $B(C_6F_5)_2(C_6Cl_5)$ and 2,2,6,6-tetramethylpiperidine with dihydrogen. Characterization of the salt $[C_5H_6Me_4NH_2][HB(C_6F_5)_2(C_6Cl_5)]$ reveals a close approach of the hydride and proton on boron and nitrogen of 1.8047(12) Å. This interaction is non-linear and suggests B–H$\cdots$H–N dihydrogen bond linking the ion pair. More recently, Holbrey, Swadźba-Kwaśny, and coworkers described neutron diffraction experiments as well as NMR studies using eutectic solvents for the FLP t-Bu$_3$P/B(C$_6$F$_5$)$_3$.[27] Correlation analysis of the neutron diffraction data supported the presence of the weakly-associated encounter

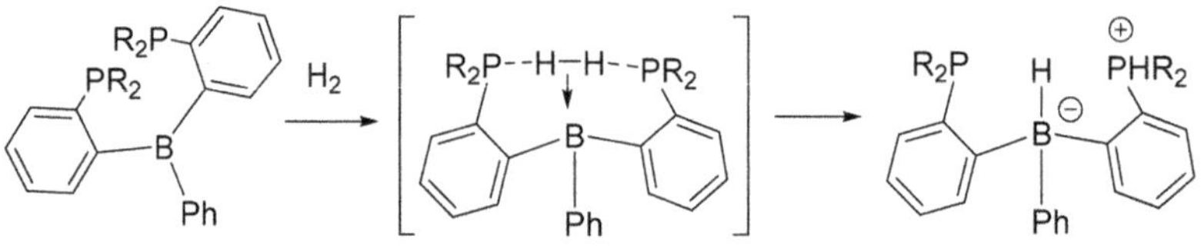

Scheme 2.3   Proposed η^2-interaction of dihydrogen with boron, stabilized by phosphine donors.

complex, accounting for about 5% of the FLP in benzene solution. In the ionic liquid, 1-decyl-3-methylimidazolium bis-triflamide, NMR data suggest the formation of the encounter complex which constituted about 20% of the dissolved species.

Autrey *et al.*[28] determined thermodynamic and kinetic parameters for FLP hydrogen activation using $Mes_2PCH_2CH_2B(C_6F_5)_2$ by reaction calorimetry. The reaction enthalpy for the conversion to the zwitterion $Mes_2PHCH_2CH_2BH(C_6F_5)_2$ was determined to be $\Delta H_r = -7.5 \pm 1.0$ kcal mol^{-1}. Moreover, they confirmed that dihydrogen activation is rate-determining in the catalysis of imine hydrogenation. An overall second-order rate constant for this reaction was determined to be $k_r = 0.9$ M^{-1} s^{-1} at 295 K.[28]

In a later study, Bourissou, Szieberth, and coworkers[29] probed the interaction of dihydrogen with $PhB(C_6H_4Pi\text{-}Pr_2)_2$. These authors identified an intermediate species computationally that involved an η^2-interaction of dihydrogen with the boron concurrent with sigma donation from both phosphines *en route* to the product $PhBH(C_6H_4Pi\text{-}Pr_2)$ $(C_6H_4PHi\text{-}Pr_2)$ (Scheme 2.3).

2.6   Radical Pathway?

Also worthy of consideration is the possibility that FLPs operate *via* a radical mechanism.[30] Combination of $t\text{-}Bu_3P$ and the Lewis acids $E(C_6F_5)_3$ (E = Al, B) were shown to react with Ph_3SnH effecting the expected heterolytic Sn–H bond cleavage providing the products $[t\text{-}Bu_3PSnPh_3]X$ (X = $[HB(C_6F_5)_3]$, $[(\mu\text{-}H)(Al(C_6F_5)_3)_2]$). In marked contrast, the corresponding reactions using Mes_3P afforded the products $[Mes_3PH]X$ (X = $[HB(C_6F_5)_3]$, $[(\mu\text{-}H)(Al(C_6F_5)_3)_2]$) and $Ph_3SnSnPh_3$. In probing these differing reaction pathways, the combination of Mes_3P and $E(C_6F_5)_3$ (E = B, Al) were shown to generate EPR-active species, inferring single electron transfer, and thus, transient generation of the 'frustrated radical pair' (FRP): $[Mes_3P]^{\bullet+}/[E(C_6F_5)_3]^{\bullet-}$. Each of these radicals was proposed to react with Ph_3SnH to abstract H• affording the observed salts, while dimerization of the product Sn radical affords $Ph_3SnSnPh_3$ (Scheme 2.4).[30] The analogous reaction pathway cannot be unambiguously shown for dihydrogen, as the products of heterolytic and homolytic H–H cleavage are identical. Nonetheless, these observations suggest that radical pathways may be operative for dihydrogen activation by some FLPs. Interestingly, a closely related radical pathway has also been recently proposed for reactions of silylium ions $[R_3Si]^+$ and $PMes_3$.[31]

While these results infer a radical mechanism for this reaction of Ph_3SnH, Slootweg and coworkers[32] subsequently showed the rates of reaction of R_3P, $B(C_6F_5)_3$ and Ph_3SnH or dihydrogen are not accelerated by photolysis which is known to prompt electron

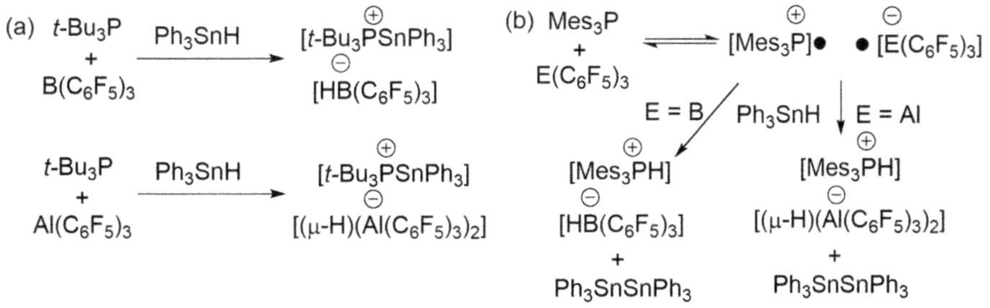

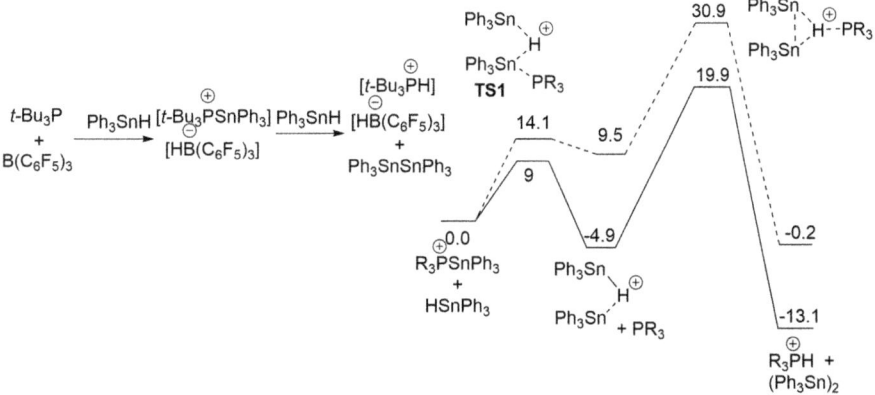

Scheme 2.4 Initial evidence suggesting (a) heterolytic and (b) radical mechanisms for the combination of R_3P (R = t-Bu, Mes) and $E(C_6F_5)_3$ (E = B, Al).

Scheme 2.5 Subsequent study shows the subsequent reaction of P/B FLPs with $HSnPh_3$. The computed kinetic barriers to the phosphonium cations for R_3P (R = Mes, t-Bu).

transfer and the generation of the radical pair, $[Mes_3P]^{+\bullet}/[B(C_6F_5)_3]^{-\bullet}$. Moreover, in the presence of excess Ph_3SnH, the conversion of $[R_3PSnPh_3][HB(C_6F_5)_3]$ to $[R_3PH]$ $[HB(C_6F_5)_3]$ as well as computational data suggest that reaction of initially formed heterolytic cleavage product $[Mes_3PSnPh_3][HB(C_6F_5)_3]$, which reacts with an additional equivalent of Ph_3SnH to give $[Mes_3PH][HB(C_6F_5)_3]$ and $Ph_3SnSnPh_3$, is energetically favorable for $PMes_3$ but is kinetically disfavored in the case of t-Bu$_3P$ (Scheme 2.5). It is interesting to note that while these latter results suggest that a radical mechanism is unlikely for the FLP activation of dihydrogen with P/B FLPs, Slootweg and coworkers[33] have also described the reactions of tris(3,5-dinitromesityl)borane and tris(mesityl)borane with dihydrogen in the presence of either Na or Cp^*_2Co as a chemical reductant provides a radical pathway to $[Cp^*_2Co][HBAr_3]$.

2.7 Intermolecular FLPs: variations of the Lewis Base

The question of the generality of the ability of Lewis acid-base combinations to affect dihydrogen activation has been probed in many studies. In this section, we focus on those in which the Lewis base component of intermolecular FLPs was varied.

Initial variations focused on the facile modifications of the base used in combination with $B(C_6F_5)_3$ and the resulting reactivity of the FLP with dihydrogen. A 2008 study from the Erker group demonstrated that the combination of 1,8-bis(diphenylphosphino)-naphthalene and $B(C_6F_5)_3$ acted as an intermolecular FLP to heterolytically cleave dihydrogen under mild conditions (1.5 bar), affording the salt $[C_{10}H_6(PPh_2)_2H][HB(C_6F_5)_3]$ (Scheme 2.6).[34] The single proton in the cation was shown to undergo rapid exchange between the two phosphine sites, although this could be slowed at low temperatures. Interestingly, the crystallographic data for this salt showed the close approach of the P–H of the phosphonium cation and the B–H of the hydridoborate anion $[HB(C_6F_5)_3]$ at 2.08 Å. This salt was shown to liberate dihydrogen at 60 °C, regenerating the FLP system. Accordingly, this represented another example of fully reversible metal-free activation of dihydrogen, indicating that the initial FLP system was not unique.

Another early extension of FLP reactivity involved the use of nitrogen-based donors in place of phosphorus bases. Thus, the stoichiometric combination of the sterically encumbered ketimine $(i\text{-}Pr_2C_6H_3)N=CMe(t\text{-}Bu)$ with $B(C_6F_5)_3$ reacts with dihydrogen to give the iminium salt $[(i\text{-}Pr_2C_6H_3)N(H)=CMe(t\text{-}Bu)][HB(C_6F_5)_3]$ (Scheme 2.7).[35]

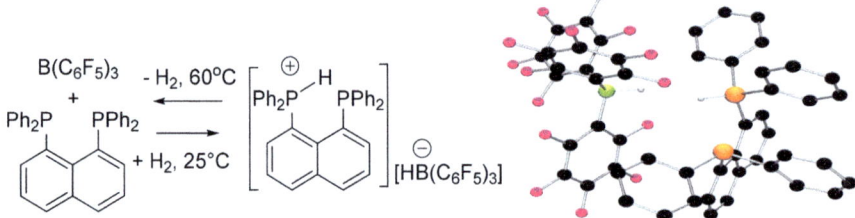

Scheme 2.6   Reversible dihydrogen activation by $C_{10}H_6(PPh_2)_2/B(C_6F_5)_3$ and the structure of the salt $[C_{10}H_6(PPh_2)_2H][HB(C_6F_5)_3]$.

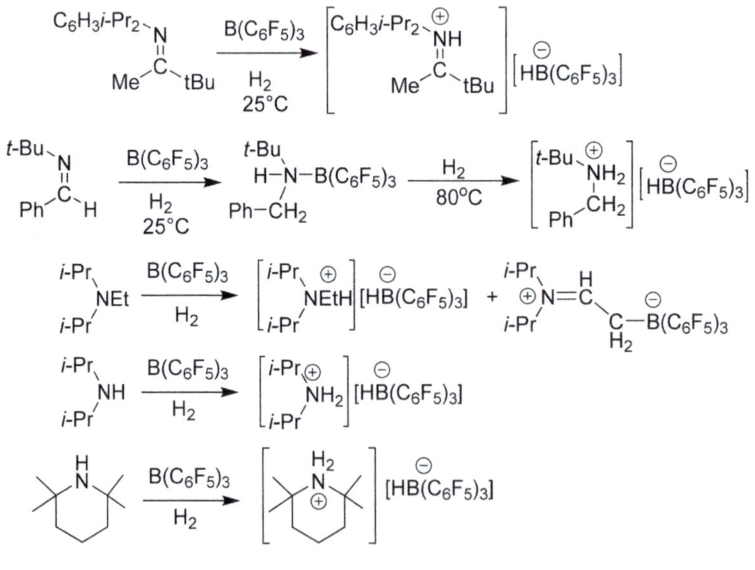

Scheme 2.7   Activation of dihydrogen by imines and amines with $B(C_6F_5)_3$.

In a related reaction, the combination of the imine t-BuN=CPh(H) and $B(C_6F_5)_3$ also reacts with dihydrogen, leading to the formal reduction of the imine and formation of the amine–borane adduct, t-Bu(PhCH$_2$)NH(B(C$_6$F$_5$)$_3$) (Scheme 2.7). This observation infers the transient formation of an iminium hydridoborate [t-BuNH=CPh(H)][HB(C$_6$F$_5$)$_3$], which is unstable resulting in hydride transfer to the iminium carbon. Subsequent heating of t-Bu(PhCH$_2$)NH(B(C$_6$F$_5$)$_3$) at 80 °C for 1 h under 4–5 bar of dihydrogen resulted in a further reaction, providing the ammonium hydridoborate salt, [t-BuNH$_2$(CH$_2$Ph)][HB(C$_6$F$_5$)$_3$] (Scheme 2.7).[35] This finding was further exploited to develop FLP hydrogenation catalysts (*vide infra*, Chapter 3).

Similarly, the reactivity of combinations of the amine bases i-Pr$_2$NEt and i-Pr$_2$NH with $B(C_6F_5)_3$ was also explored. These pairs do not afford the corresponding classical Lewis adducts. Instead, the known ability of the Lewis acid $B(C_6F_5)_3$ to effect the dehydrogenation of amines with *beta*-hydrogen atoms results in 50:50 mixtures of the corresponding ammonium salts, [i-Pr$_2$NHR][HB(C$_6$F$_5$)$_3$] (R = Et, H), and the zwitterionic products, i-Pr$_2$N=CHCH$_2$B(C$_6$F$_5$)$_3$ and i-PrNH=CMe(CH$_2$)B(C$_6$F$_5$)$_3$ (Scheme 2.7).[36] Regardless, the addition of dihydrogen to mixtures of i-Pr$_2$NH or Me$_4$C$_5$H$_6$NH with $B(C_6F_5)_3$ gave the quantitative formation of the ammonium hydridoborate salts, [i-Pr$_2$NH$_2$][HB(C$_6$F$_5$)$_3$] and [Me$_4$C$_5$H$_6$NH$_2$][HB(C$_6$F$_5$)$_3$], respectively (Scheme 2.7).[36] Corresponding efforts to activate dihydrogen with silyl amines such as i-Pr$_2$NSiMe$_3$ and $B(C_6F_5)_3$ proceeded but rapidly led to the loss of the silane, Me$_3$SiH, affording i-Pr$_2$NH.[37]

In our efforts, we exploited ferrocenyl phosphine derivatives in FLP chemistry.[38] The mono- and bis-ferrocenyl phosphines (η^5-C$_5$H$_4$Pt-Bu$_2$)FeCp and (η^5-C$_5$H$_4$PR$_2$)$_2$Fe (R = i-Pr, t-Bu) were combined with $B(C_6F_5)_3$ resulting in the formation of the *para*-attack products (η^5-C$_5$H$_4$Pt-Bu$_2$C$_6$F$_4$BF(C$_6$F$_5$)$_2$)FeCp, (η^5-C$_5$H$_4$Pt-Bu$_2$C$_6$F$_4$BF(C$_6$F$_5$)$_2$)Fe(η^5-C$_5$H$_4$Pt-Bu$_2$) and (η^5-C$_5$H$_4$Pi-Pr$_2$C$_6$F$_4$BF(C$_6$F$_5$)$_2$)$_2$Fe, respectively (Scheme 2.8). Presumably, in the case of (η^5-C$_5$H$_4$Pt-Bu$_2$C$_6$F$_4$BF(C$_6$F$_5$)$_2$)Fe(η^5-C$_5$H$_4$Pt-Bu$_2$), the product is sufficiently sterically encumbered, thus precluding a second reaction. However, the subsequent reaction between (η^5-C$_5$H$_4$Pt-Bu$_2$C$_6$F$_4$BF(C$_6$F$_5$)$_2$)Fe(η^5-C$_5$H$_4$Pt-Bu$_2$) and chlorodimethylsilane afforded (η^5-C$_5$H$_4$Pt-Bu$_2$C$_6$F$_4$BH(C$_6$F$_5$)$_2$)Fe(η^5-C$_5$H$_4$Pt-Bu$_2$), which behaves as a sterically encumbered phosphine, forming an FLP upon addition of $B(C_6F_5)_3$. This FLP system effects the heterolytic activation of dihydrogen (4 bar) to give the unusual salt [(η^5-C$_5$H$_4$Pt-Bu$_2$C$_6$F$_4$BH(C$_6$F$_5$)$_2$)Fe(η^5-C$_5$H$_4$PHt-Bu$_2$)][HB(C$_6$F$_5$)$_3$] (Scheme 2.8). The closely related and more sterically encumbered ferrocenyl phosphine, (η^5-C$_5$H$_4$Pt-Bu$_2$)Fe(C$_5$Ph$_5$), also forms an FLP upon addition of $B(C_6F_5)_3$, showing no evidence of the *para*-substitution reaction. Furthermore, upon exposure of 4 bar of dihydrogen to this mixture, the salt [(η^5-C$_5$H$_4$PHt-Bu$_2$)Fe(C$_5$Ph$_5$)][HB(C$_6$F$_5$)$_3$] is formed (Scheme 2.8).

In a closely related work, the Erker group[39] examined the reaction of a ferrocene derivative with a phosphine substituent on an alkyl chain linked to one of the cyclopentadienyl rings, (C$_5$H$_4$)$_2$(CHMeCH$_2$CHPMes$_2$)Fe. In combination with $B(C_6F_5)_3$ and dihydrogen, this led to an intermediate resulting from the heterolytic cleavage of dihydrogen; however, the transient cation and anion react further, cleaving the P–C bond, affording (C$_5$H$_4$)$_2$(CHMeCH$_2$CH$_2$)Fe and (C$_6$F$_5$)$_3$BPHMes$_2$ (Scheme 2.8).

The related zirconocene-dichloride derivative, (η^5-C$_5$H$_4$CH$_2$NH(C$_6$H$_3$$i$-Pr$_2$))$_2$ZrCl$_2$ was also shown to act as the Lewis basic component of an FLP. The combination with $B(C_6F_5)_3$ effected the FLP activation of dihydrogen to afford the mono- and

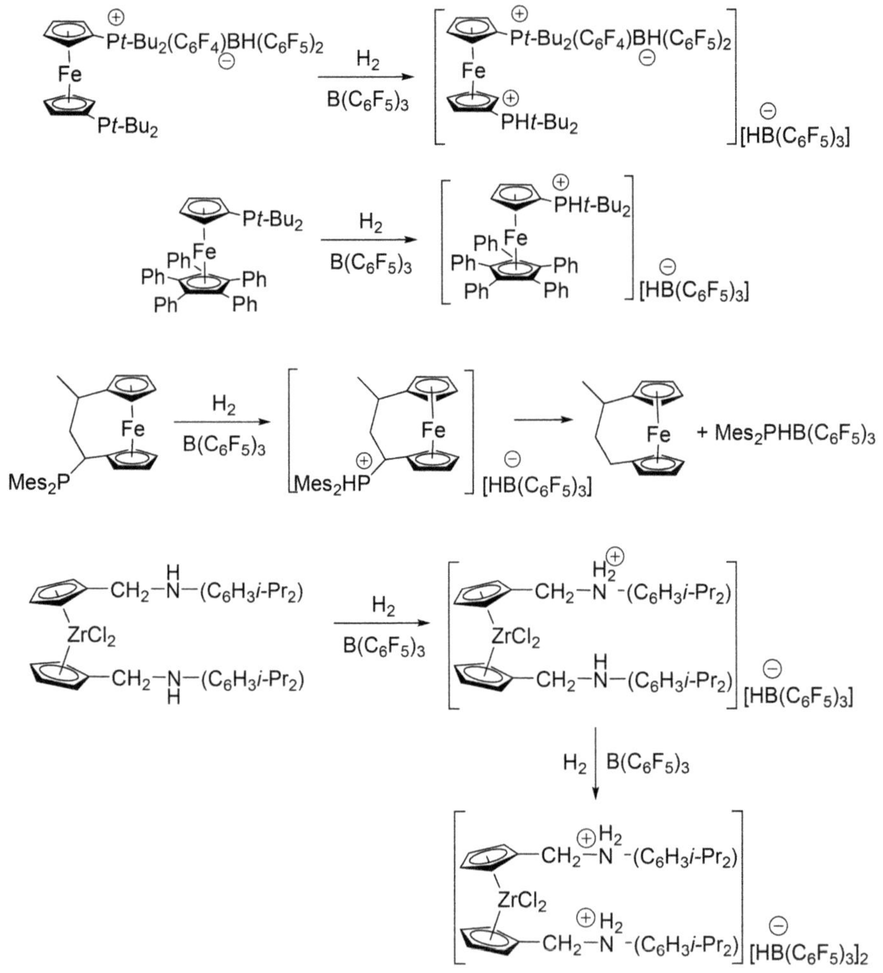

Scheme 2.8 Reactions of metallocene phosphines/B(C$_6$F$_5$)$_3$ with dihydrogen.

bis-ammonium salts, $[(\eta^5\text{-}C_5H_4CH_2NH(C_6H_3i\text{-}Pr_2))(\eta^5\text{-}C_5H_4CH_2NH_2(C_6H_3i\text{-}Pr_2)ZrCl_2]$ $[HB(C_6F_5)_3]$ and $[(\eta^5\text{-}C_5H_4CH_2NH_2(C_6H_3i\text{-}Pr_2))_2ZrCl_2][HB(C_6F_5)_3]_2$, in consecutive reactions. (Scheme 2.8).[40,41] In a subsequent study, Tamm and coworkers[42] reported the analogous reaction of $(COT)TiC_5H_4Pt\text{-}Bu_2$ and $B(C_6F_5)_3$ with dihydrogen, affording $[(COT)TiC_5H_4PHt\text{-}Bu_2][HB(C_6F_5)_3]$. The above systems based on organometallic backbones are novel examples in which the metal centers perform a function that is ancillary to the chemistry that occurs at the pendant ligands. This unusual situation 'turns the tables' on the traditional combination of transition metals and main group elements.

The impact of alteration of the substituents on phosphorus donors on the FLP dihydrogen activation was probed. Even though $P(O\text{-}2,4\text{-}(t\text{-}Bu_2C_6H_3))_3$ and $P(O\text{-}2,6\text{-}Me_2C_6H_3)_3$ generate FLPs with $B(C_6F_5)_3$, these combinations do not react with dihydrogen.[43] Similarly, the more basic species $t\text{-}Bu_2POR$, $(R = Ph; 2,6\text{-}Me_2C_6H_3)$ and $t\text{-}Bu_2PCl$ also form FLPs with $B(C_6F_5)_3$ but fail to react with dihydrogen. In contrast, bulky alkyl–aryl phosphines of the form $R_2PC_6H_4(2,4,6\text{-}i\text{-}Pr_3C_6H_2)$ (R = *t*-Bu, Cy) in combination with $B(C_6F_5)_3$ activated dihydrogen to give the corresponding phosphonium–hydridoborates.[44] Analogous reactivity was also demonstrated for combinations of the phosphinimines Ph_3PNR (R = Ph, *t*-Bu, C_6F_5) and $B(C_6F_5)_3$ with dihydrogen afforded the phosphinammonium salts $[Ph_3PN(H)R][HB(C_6F_5)_3]$,[45] while a very recent study[46] extended such FLP activation of dihydrogen to aminophosphines in combination with $B(C_6F_5)_3$.

In another variation,[47] a bis-phosphine derived calix[4]arene $[C_6H_2(OPr)(PMes_2)$ $CH_2C_6H_3(OPr)CH_2]_2$ has been prepared and shown to react with two equivalents of dihydrogen in the presence of $B(C_6F_5)_3$ affording the salt $[C_6H_2(OPr)(PHMes_2)CH_2C_6H_3(OPr)$ $CH_2]_2[HB(C_6F_5)_3]_2$ (Scheme 2.9).

While pyridine forms an adduct with $B(C_6F_5)_3$,[13,14] the work of H. C. Brown *et al.*[15] illustrated that the steric demands of lutidine could inhibit interactions with boron Lewis acids (see Section 1.8). This prompted the examination of the reaction of 2,6-lutidine and $B(C_6F_5)_3$ in the context of FLP chemistry.[48] The mixture of these reagents gave rise to broad NMR spectra, inferring the presence of an equilibrium mixture between the free reagents and the Lewis acid–base adduct $(2,6\text{-}Me_2C_5H_3N)$ $B(C_6F_5)_3$ (Scheme 2.9). The equilibrium constants, ΔH and ΔS, were found to be −42(1) kJ mol^{-1} and −131(5) J mol^{-1} K^{-1}, respectively, by variable-temperature NMR spectroscopy. The low temperature ^{19}F NMR data were also consistent with the formation of the adduct, which was isolated from solutions at −40 °C. The steric congestion in this adduct was evident from the boron–nitrogen bond length (1.661(2) Å), which is significantly longer than that in $(py)B(C_6F_5)_3$ (1.628(2) Å).[13,14] Subsequent treatment of the room temperature equilibrium mixture with dihydrogen, gave the lutidinium salt $[2,6\text{-}Me_2C_5H_3NH][HB(C_6F_5)_3]$ (Scheme 2.10). This observation reveals that despite the formation of a classical Lewis adduct, the equilibrium provides access to the FLP. This view was also demonstrated for related pyridines with similar steric encumbrance,[49] while more sterically demanding pyridines and amines precluded formation of classical Lewis acid–base adducts completely, leading solely to FLP reactivity with dihydrogen.[50] Interestingly the combination of lutidine and BCl_3 was shown to dissociate at elevated temperatures and react with dihydrogen, although chloride/ hydride redistribution affords $(C_5H_3Me_2N)BHCl_2$ and $[C_5H_3Me_2NH][BCl_4]$.[51] A calorimetric study[52] examined the thermodynamics of dihydrogen activation by lutidine $B(C_6F_5)_3$, while a subsequent computational study supported the notion of an encounter complex analogous to that seen for *t*-Bu$_3$P/B(C_6F_5)_3.[53] In a separate study,

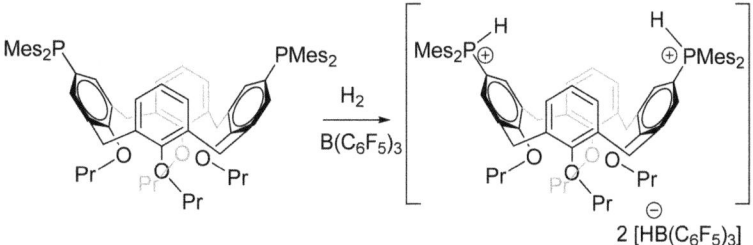

Scheme 2.9 Calixarene derived bis-phosphine in FLP activation of dihydrogen.

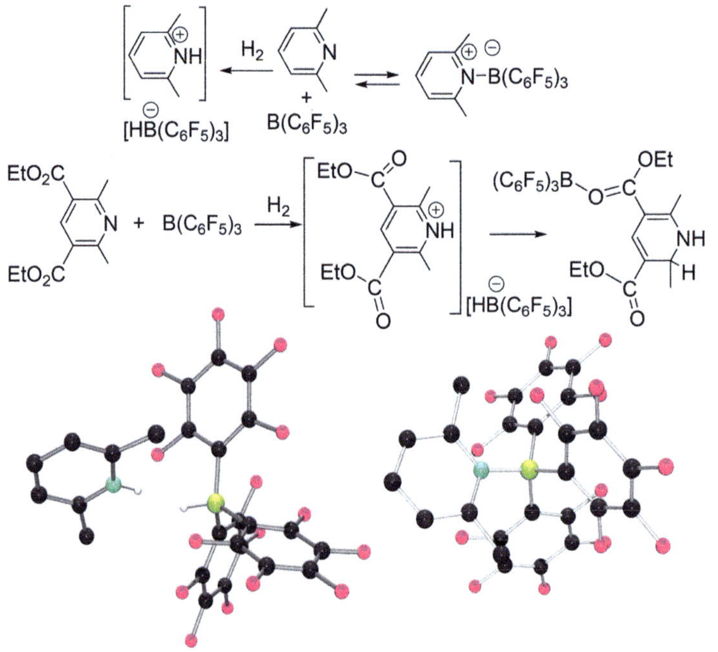

Scheme 2.10   Classical and FLP reactivity of lutidine or Hantzsch's ester with $B(C_6F_5)_3$.

Autrey *et al.*[54] have computationally examined the impact of solvent effects on the formation of FLPs. A further collaborative study with the Crudden group,[55] revealed that a Hantzsch's ester, derived from lutidine, also reacted with $B(C_6F_5)_3$ and dihydrogen to give $[C_5H(CO_2Et)_2Me_2NH][HB(C_6F_5)_3]$ although this goes on to reduce the pyridinium *via* the addition of hydride to the 2-carbon and coordination of borane to the ester oxygen (Scheme 2.10).

   These findings emphasize that classical Lewis adducts and FLP reactivity are not mutually exclusive, but rather are a continuum of reactivity. Furthermore, it suggests the possibility that classical Lewis acid-base adducts could dissociate, and thus be reactive. It is interesting to note that because such adducts have been viewed as thermodynamic sinks, their reactivity has been largely overlooked. This aspect is discussed further in Section 2.11.

   A unique strategy[56] to the sterically encumbered nitrogen donors involved the use of supermolecular interactions. While the amine $(3,5-C_6H_3)NH(CH_23,5-C_6H_3)$ forms a classical Lewis acid–base adduct with $B(C_6F_5)_3$, which does not react with dihydrogen, the analogous rotaxane, $(3,5-C_6H_3)NH(CH_23,5-C_6H_3)$ @ $(CH_2CH_2O)_6(CH_2)_x$ $(X = 4, 6)$ in the presence of $B(C_6F_5)_3$ behaves as an FLP to activate dihydrogen, affording the corresponding ammonium cation encircled by the crown with the $[HB(C_6F_5)_3]$ anion (Scheme 2.11). This finding indicates that steric encumbrance sufficient to induce FLP behavior can be achieved with mechanically interlocked molecules. A subsequent computational study inferred that the hydrogen bonding in the product between the ammonium and the crown has a profound effect on the dihydrogen activation.[57]

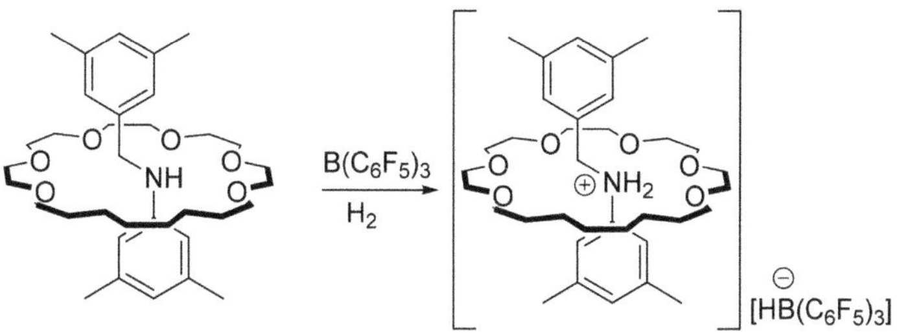

Scheme 2.11 A rotaxane-derived intermolecular N/B FLP, reaction with dihydrogen.

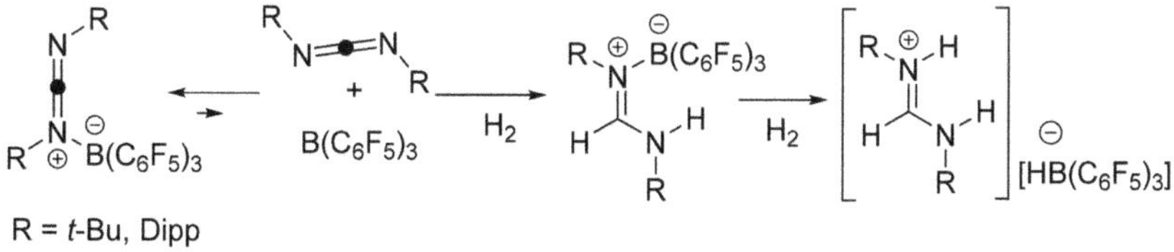

R = *t*-Bu, Dipp

Scheme 2.12 Intermolecular FLP derived from carbodiimide adduct.

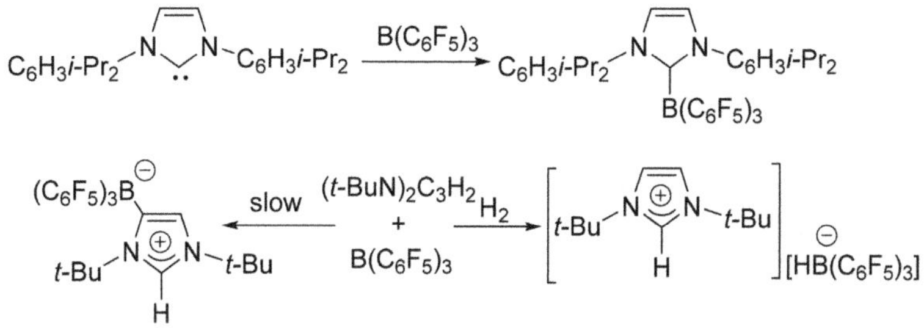

Scheme 2.13 Activation of dihydrogen involving carbenes.

Another intermolecular B/N FLP was derived from the adduct formed by the combination of a carbodiimide and $B(C_6F_5)_3$.[58] The resulting adduct reacts with two equivalents of dihydrogen in a sequential fashion, affording access to $RN(H)C(H)NRB(C_6F_5)_3$ and $[(RN(H))_2CH][HB(C_6F_5)_3]$ (Scheme 2.12).

Beyond N and P donors, the examination of carbenes in FLP chemistry was prompted by the broad use of N-heterocyclic carbenes in organometallic chemistry. This was an interesting possibility given that N-heterocyclic carbenes do not react with dihydrogen on their own. This issue was addressed in simultaneous studies from the Stephan[59,60] and Tamm[61–64] groups. While the carbene $((C_6H_3i\text{-}Pr_2)N)_2C_3H_2$ was shown to form a stable, classical Lewis acid–base adduct with $B(C_6F_5)_3$ (Scheme 2.13), the related carbene $(t\text{-}BuN)_2C_3H_2$ and $B(C_6F_5)_3$ generated an FLP. Tamm *et al.*[61] showed that this mixture eventually affords the abnormal-carbene–borane adduct on prolonged standing (Scheme 2.13). However, exposure to dihydrogen resulted in heterolytic cleavage of dihydrogen affording the imidazolium hydridoborate salt, $[(t\text{-}BuN)_2C_3H_3][HB(C_6F_5)_3]$ (Scheme 2.13).

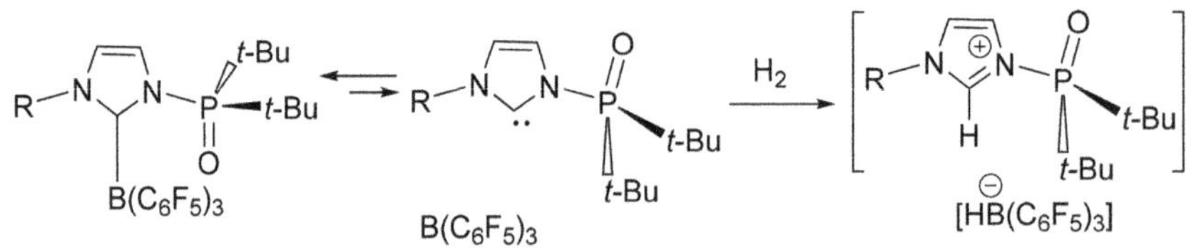

Scheme 2.14   Rotational control affording access to FLP.

It is noteworthy that Tamm *et al.* also described an 'encounter-complex' for the transition state involving the activation of dihydrogen by carbene $(t\text{-BuN})_2C_3H_2$ and the borane $B(C_6F_5)_3$. Tamm also showed that the FLP derived from the borane $B(3,5\text{-}(CF_3)_2C_6H_3)_3$ and the carbene $(t\text{-BuN})_2C_3H_2$ affected dihydrogen activation, while the analog reaction using $((C_6H_3i\text{-Pr}_2)N)_2C_3H_2$ formed a stable non-reactive Lewis acid–base adduct.[64]

In an elegant tuning of FLP reactivity, Hoshimoto and coworkers[65] designed carbenes of the form $(RN)(t\text{-Bu}_2P(O)N)C_3H_2$, $(R = C_6H_3i\text{-Pr}_2, \text{Mes}, 3,5\text{-}t\text{-Bu}_2C_6H_3)$. These species exhibit inhibited rotation about the P–N bond. In the orientation where the PO bond is oriented on the same side as the carbene, $B(C_6F_5)_3$ binding to the carbene is sterically allowed, whereas this binding is frustrated when the PO is oriented towards the backside of the carbene. This allows temperature-controlled access to the FLP in temperatures ranging from 60–120 °C depending on the steric demands of the R group on nitrogen (Scheme 2.14). This also allowed both the reaction with dihydrogen affording the imidazolium hydroborate and the stoichiometric hydrogenation of $PhC(H)=NSO_2Ph$ at elevated temperature.

2.8   Intermolecular FLPs: variations of the Lewis Acid

Another approach to variation of the nature of intermolecular FLPs is to consider the Lewis acid component. In general, this has received lesser initial attention, as suitable sterically encumbered Lewis acids are less commonly available. Nonetheless, some studies have probed the impact of such variation. As mentioned in Chapter 1, the reaction of $t\text{-Bu}_3P$ and BPh_3 with dihydrogen gave $[t\text{-Bu}_3PH][HBPh_3]$ albeit in low yield, while the corresponding reaction of Mes_3P and BPh_3, or $t\text{-Bu}_3P$ and $BMes_3$ showed no reaction.[66] In a similar fashion, Repo and Rieger reported that using amines as the base and BPh_3 as the Lewis acid also led to no reactions with dihydrogen.[36]

Targeting a borane that was sufficiently Lewis acidic to effect dihydrogen activation in conjunction with a Lewis base, and yet suitably modified to preclude *para*-attack by phosphines, we targeted a seemingly trivial modification of the borane. The species $B(p\text{-}C_6F_4H)_3$ was prepared *via* a synthetic protocol involving treatment of the $BF_3{\cdot}OEt_2$ with the corresponding aryl Grignard reagent.[67] While the initial product was isolated as the diethyl ether adduct, sublimation under reduced pressures afforded the base-free borane. In combination with sterically encumbered phosphines (*i.e.*, PR_3, where $R = t\text{-Bu}, Cy, o\text{-}C_6H_4Me$), the corresponding FLPs were generated with no evidence of unproductive side reactions. However, upon exposure to dihydrogen at 25 °C, the corresponding phosphonium hydridoborate salts were formed[67,68] (Scheme 2.15). In the case of $[(o\text{-}C_6H_4Me)_3PH][HB(p\text{-}C_6F_4H)_3]$, the example with the weakest conjugate base, this species was found to slowly evolve dihydrogen under vacuum at 25 °C (85% consumption after 9 days). At 80 °C, this evolution was accelerated and complete conversion of the salt to the FLP was observed after 12 h.[67] A subsequent detailed computational study by Privalov and coworkers[69] probed the details of both the activation and release of dihydrogen by this system. These results further support the notion that there is a 'sweet-spot' of combined Lewis acidity and basicity for FLPs. This view further infers that the Lewis acid and base components can be judiciously tuned to optimize the reversible nature of dihydrogen activation and release. Interestingly, use of the bulky phosphine $(Me_3Si)_3P$ with $B(C_6F_4H)_3$ led to the slow activation of dihydrogen followed by a fast redistribution reaction affording $[(Me_3Si)_4P][HB(C_6F_4H)_3]$ and the phosphine–borane adduct, $((Me_3Si)_2HP)B(C_6F_4H)_3$ (Scheme 2.15).[70]

Another study employing Lewis acid variants was reported by Berke et al.[71] These authors used tetramethylpiperidine with the boranes $RB(C_6F_5)_2$ (R = Cy, PhC_2H_4) derived from the hydroboration of precursor olefins. While the parent system with R = C_6F_5 gives an irreversible dihydrogen activation, using the slightly less Lewis acidic boranes, $RB(C_6F_5)_2$ results in reversible dihydrogen-splitting at elevated temperature.

Other variations of boron-based Lewis acids have also been explored in FLP activation of dihydrogen. For example, the O'Hare group used tris(2,2',2''-perfluorobiphenyl) borane[72] and Ashley et al.[73] used $(3,5\text{-}(CF_3)_2C_6H_2)_3B$ both in combination with tetramethylpiperidine for dihydrogen activation. In a similar sense, Wang, and coworkers[74] reported the related bis[[(2,4,6-trifluoromethyl)phenyl]borane which, in combination with the Lewis base $(CH_2CH_2)_3N_2$, reacts with dihydrogen to give the salt $[C_6H_{13}N_2][((CF_3)_3C_6H_2)_2BH_2]$ (Scheme 2.16). It is also noteworthy that the structural isomers $(2,5\text{-}(CF_3)_2C_6H_2)_3B$ and $(2,4\text{-}(CF_3)_2C_6H_2)_3B$ did not react with dihydrogen in the

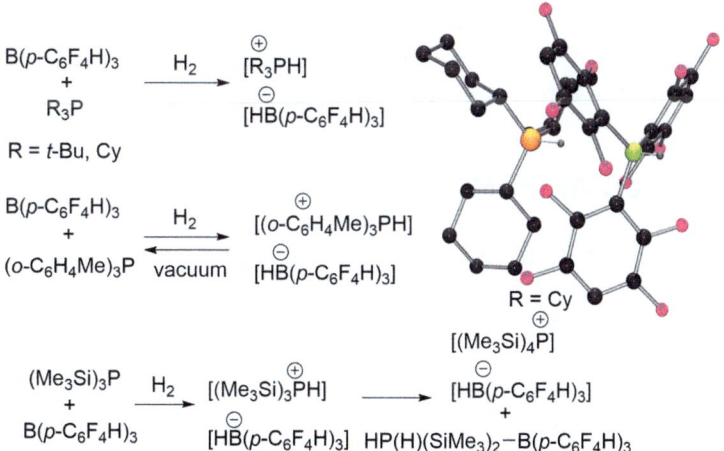

Scheme 2.15 Reactions of $B(p\text{-}C_6F_4H)_3$ with phosphines and dihydrogen and structure of $[Cy_3PH][BH(p\text{-}C_6F_4H)_3]$.

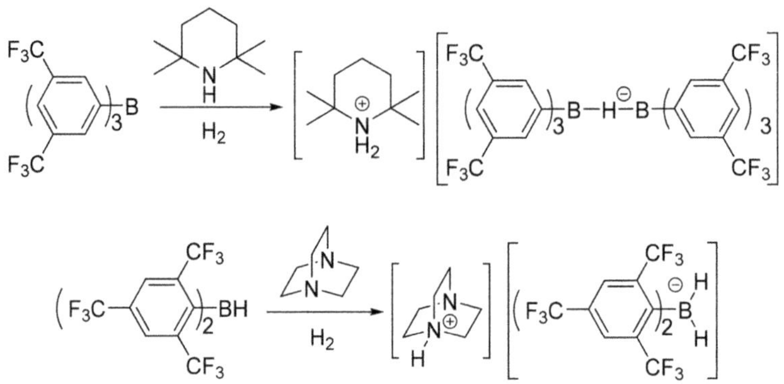

Scheme 2.16 Boron Lewis acid variations in FLP dihydrogen activation.

presence of $C_5H_6Me_4NH$, presumably a result of the steric hindrance of the ortho-CF_3 fragments.[75] In contrast, the species $(3,5-(CF_3)_2C_6H_2)(C_6X_5)_2B$, $(3,5-(CF_3)_2C_6H_2)_2(C_6X_5)B$, $(X = F, Cl)$, $(C_6F_5)(C_6Cl_5)_2B$ and $(C_6F_5)_2(C_6Cl_5)B$ activated dihydrogen in the present of t-Bu_3P affording the corresponding phosphonium hydridoborate salts.[76] Similarly the borane $(C_6F_5)(C_6Cl_5)(3,5-(CF_3)_2C_6H_2)B$ activated dihydrogen in the presence of lutidine, $C_5H_6Me_4NH$ or t-Bu_3P.[77]

Another alteration of the Lewis acids involves the use of other boron derivatives. For example, the mixture of $B(OC_6F_5)_3$ and Pt-Bu_3 generates an FLP, however on reaction with dihydrogen the salt $[HPt$-$Bu_3][B(OC_6F_5)_4]$ is formed, presumably resulting from sequential dihydrogen activation and borate-substituent redistribution. Efforts to use the boronic esters $(C_6H_4O)_2BC_6F_5$, $(C_6H_3FO)_2BC_6F_5$ and $(C_6F_4O)_2BC_6F_5$, and borate esters $B(OC_6H_3(CF_3)_2)_3$, $B(OC_6H_2F_3)_3$ and $B(OC_6H_4CF_3)_3$, in FLPs with t-Bu_3P or Mes_3P led to no reaction with dihydrogen.[43] In contrast, the borane $B(CH(C_6F_5)_2)_3$ in combination with the relatively strong base $C_5H_6Me_4NH$, affects dihydrogen activation, affording the expected ammonium hydridoborate, although this reaction was slow requiring 4 days at 90 °C and 4 bar of dihydrogen pressure.[78] Collectively these data further support the initially proposed notion[66] that there is a threshold of combined Lewis acidity and basicity that is required to effect the splitting of dihydrogen.

2.9   Intramolecular FLPs

Targeting intramolecular systems related to the initial $Mes_2PCH_2CH_2B(C_6F_5)_2$, the Erker group[79] prepared related systems derived from the hydroboration of alkynylphosphines. For example, the orange species, $trans$-t-$Bu_2P(CH=CMe)B(C_6F_5)_2$, was found to be inert to dihydrogen under ambient conditions (1 bar of dihydrogen, 25 °C). However, under more forcing conditions, this species reacted with dihydrogen (60 bar) to afford the corresponding zwitterionic phosphonium hydrido-borate, $[trans$-t-$Bu_2PHCH=CMeBH(C_6F_5)_2]$ (Scheme 2.17).[79] Under similar conditions, the corresponding reaction with dideuterium (D_2) provided the deuterated isotopomer. In contrast, the related species $trans$-$Mes_2P(CH=C(R))B(C_6F_5)_2$ (R = Me, Ph),

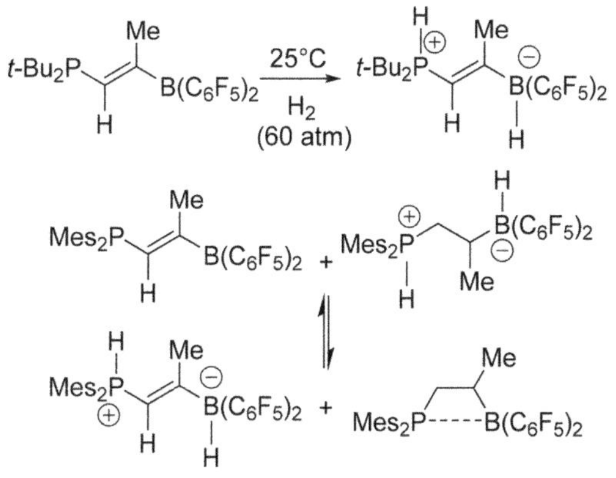

Scheme 2.17 Reactions of intramolecular alkenyl-linked phosphino–boranes.

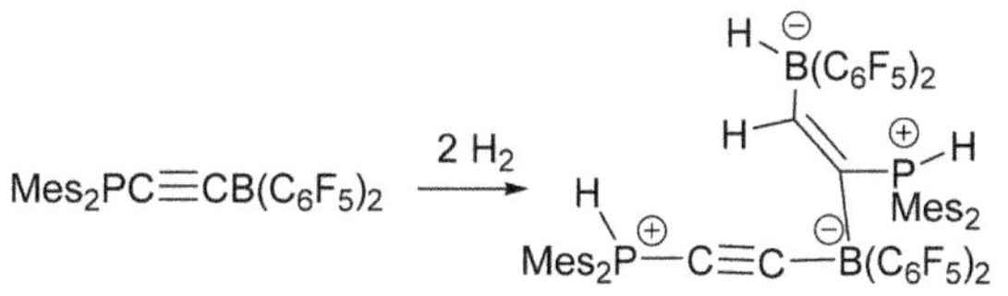

Scheme 2.18 Reaction of alkyne linked FLP with dihydrogen.

containing less basic phosphine centers, proved inert to dihydrogen even at elevated pressure (60 bar). Nonetheless, the species *trans*-Mes$_2$P(CH=C(R))B(C$_6$F$_5$)$_2$ reacted with Mes$_2$PHCH$_2$CH$_2$BH(C$_6$F$_5$)$_2$ to affect the rapid transfer of proton and hydride, thus providing an equilibrium access to the species [Mes$_2$PHCH=C(R)BH(C$_6$F$_5$)$_2$] (Scheme 2.17). Subsequent addition of 2.5 bar of dihydrogen caused a shift in the equilibrium mixture favoring the latter species. Complete conversion to olefin-linked phosphonium-hydridoborate [Mes$_2$PHCH=C(R)BH(C$_6$F$_5$)$_2$] was achieved using 10 mol% of Mes$_2$PCH$_2$CH$_2$B(C$_6$F$_5$)$_2$, under a dihydrogen atmosphere (2.5 bar). In these cases, the ethylene linked phosphine–borane, Mes$_2$PCH$_2$CH$_2$B(C$_6$F$_5$)$_2$, activates dihydrogen generating [Mes$_2$PHCH$_2$CH$_2$BH(C$_6$F$_5$)$_2$], which mediates the transfer of proton and hydride to Mes$_2$PCH=C(R)B(C$_6$F$_5$)$_2$ to regenerate Mes$_2$PCH$_2$CH$_2$B(C$_6$F$_5$)$_2$ for further dihydrogen activation. These findings suggest that the proximity of the Lewis acidic and basic sites promotes dihydrogen activation, and are consistent with the concerted and cooperative action of the Lewis acid and base. Nonetheless, a computational study[80] has suggested the possibility of an intermolecular mechanism for these systems.

The alkyne linked FLP, Mes$_2$PCCB(C$_6$F$_5$)$_2$ was prepared in a straightforward reaction of Mes$_2$PCCH and ClB(C$_6$F$_5$)$_2$. This species was shown to react with dihydrogen; however, the product was not the anticipated zwitterion, but rather, the species (Mes$_2$PH)(B(C$_6$F$_5$)$_2$CCPHMes$_2$)C=CH(BH(C$_6$F$_5$)$_2$) (Scheme 2.18) appeared to form *via* dihydrogen activation and subsequent hydroboration affording the doubly zwitterionic product.

The related P/B species $Mes_2PCH_2CH_2CH_2B(C_6F_5)_2$ was shown not to activate dihydrogen,[81] although the corresponding zwitterion could be prepared *via* an indirect route involving proton and hydride delivery. This salt readily lost dihydrogen at ambient conditions. NMR spectroscopy and calculations showed that lack of reaction with dihydrogen was not a result of hindered P$\cdots$B dissociation, but rather that the hydrogen-splitting is markedly endergonic for this species. The Erker group[82] also described the formation of the bicyclic norbornane-derived phosphine–borane, $C_7H_{10}(PMes_2)(B(C_6F_5)_2$ which rapidly reacts with dihydrogen at ambient temperature to give the corresponding zwitterionic salt.

The Erker group continued to develop multiple strategies to intramolecular B/P FLPs and their subsequent FLP chemistry. Hydroboration of alkynes provided access to olefinic intramolecular FLPs, $Mes_2PCH=C(R)B(C_6F_5)_2$ (R = $SiMe_3$, Ph)[83] which reacted with dihydrogen affording the corresponding zwitterionic phosphonium hydridoborates. More recently, Erker *et al.*[84] have reported the preparation of the phosphino–bisborane $Mes_2PCH_2CH(B(C_6F_5)_2)_2$ derived from the generation of a borata–alkene (Scheme 2.19). This species reacted with dihydrogen to afford $Mes_2P(H)CH_2CH(B(C_6F_5)_2)_2(\mu\text{-H})$ in which the hydride is bridging between the geminal boron centers. In another effort, the Erker group[85] has developed the synthesis of a unique PB FLP in which the P atom is incorporated in a five-membered ring in the species $ArPC_4H_7(B(C_6F_5)_2)$ (Scheme 2.19).

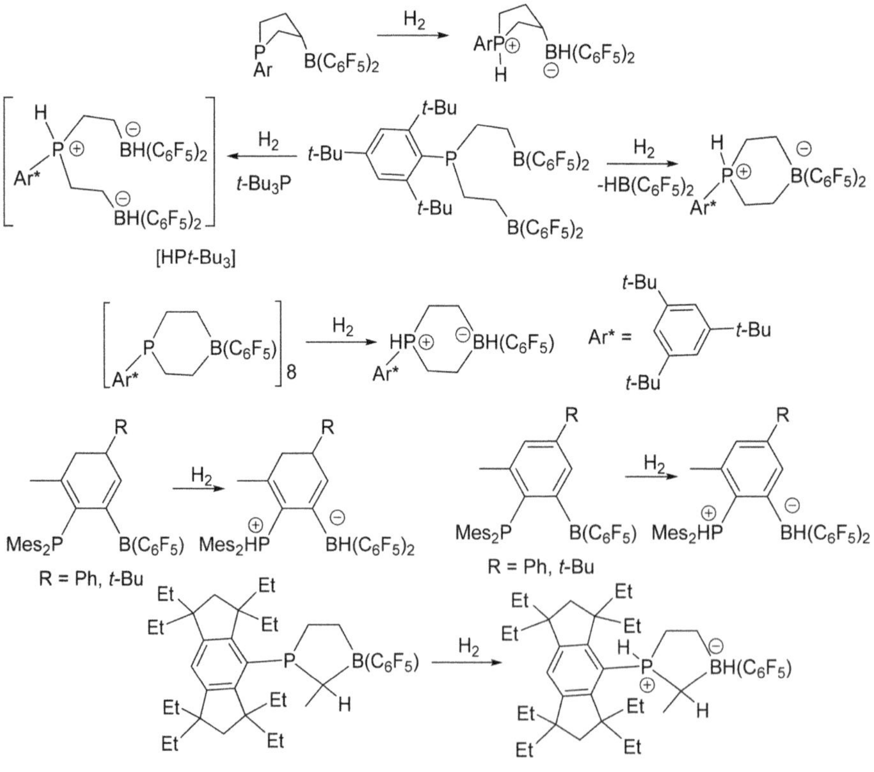

Scheme 2.19　Reactions of intramolecular FLPs with dihydrogen.

In a related sense, the FLP 2,4,6-t-Bu$_3$C$_6$H$_2$P(CH$_2$CH$_2$B(C$_6$F$_5$)$_2$)$_2$ reacted with dihydrogen and Pt-Bu$_3$ to give [2,4,6-t-Bu$_3$C$_6$H$_2$PH(CH$_2$CH$_2$BH(C$_6$F$_5$)$_2$)$_2$][HPt-Bu$_3$] (Scheme 2.19).[86] On the other hand, the direct reaction of 2,4,6-t-Bu$_3$C$_6$H$_2$P(CH$_2$CH$_2$B(C$_6$F$_5$)$_2$)$_2$ with dihydrogen proceeded with loss of HB(C$_6$F$_5$)$_2$ and the formation of 2,4,6-t-Bu$_3$C$_6$H$_2$PH(CH$_2$CH$_2$)$_2$B(C$_6$F$_5$)$_2$ (Scheme 2.19). This species reacted as expected with dihydrogen, affording the corresponding zwitterion. In related work,[87] an alternate synthetic route provided the Piers borane adduct of the cyclic 2,4,6-t-Bu$_3$C$_6$H$_2$P(CH$_2$CH$_2$)$_2$B(C$_6$F$_5$) (Scheme 2.19). Removal of the Piers' borane afforded an octameric P–B adduct, which reacted with dihydrogen to give 2,4,6-t-Bu$_3$C$_6$H$_2$PH(CH$_2$CH$_2$)$_2$BH(C$_6$F$_5$). Erker and coworkers have also used carboboration alkynyl phosphines to access olefinic and aromatic linked P/B FLPs. For example, the species (C$_6$F$_5$)$_2$BC$_6$H$_4$PhMe(PMes$_2$) and (C$_6$F$_5$)$_2$BC$_6$H$_2$PhMe(PMes$_2$) have been prepared[88,89] and shown to react with dihydrogen (Scheme 2.19). In these latter cases, solid state NMR studies confirmed the formation of the zwitterionic products.[90] Erker *et al.*[91] have their study of highly bulky substituents, using octaethylhydrindacenyl to prepare the species C$_6$H(C$_3$H$_2$Et$_4$)$_2$PCH$_2$CH$_2$B(C$_6$F$_5$)(CMeH) (Scheme 2.19). This species also reacted with dihydrogen affording the corresponding zwitterion.

In examining the design of systems targeting reversible capture of dihydrogen, Aldridge and coworkers[92] prepared the intramolecular P/B FLPs with dibenzofuran and dimethylxanthene backbones, C$_{12}$H$_6$O(PMes$_2$)(B(C$_6$F$_5$)$_2$) and C$_{13}$H$_6$Me$_2$O(PMes$_2$) (B(C$_6$F$_5$)$_2$), respectively (Scheme 2.20). Interestingly the former compound does not react with dihydrogen while the latter one does. This is attributed to the P–B distance allowing the latter species to be preorganized for the reversible activation of dihydrogen. Interestingly, Beckmann *et al.*[93] prepared the related species derived from the *peri*-substituted biphenyl, C$_{12}$H$_6$(PPh$_2$)(BMes$_2$) and showed that it failed to react with dihydrogen, presumably due to the insufficient Lewis acidity at boron.

A conceptually related intramolecular FLP exploiting a nitrogen donor incorporating tetramethylpiperidine (C$_5$H$_6$Me$_4$N)CH$_2$C$_6$H$_4$B(C$_6$F$_5$)$_2$ (Scheme 2.21), was prepared by Repo and Rieger.[94] This species reacted reversibly with dihydrogen, although the liberation of dihydrogen required heating to 120 °C. The mechanism of this reversible

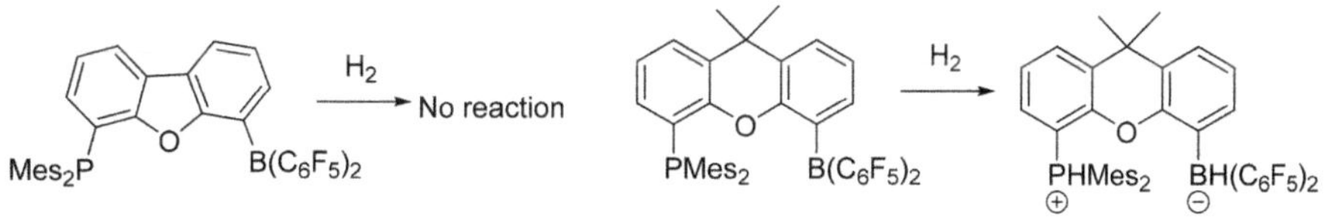

Scheme 2.20   Reactions of dibenzofuran and dimethylxanthine derived FLPs with dihydrogen.

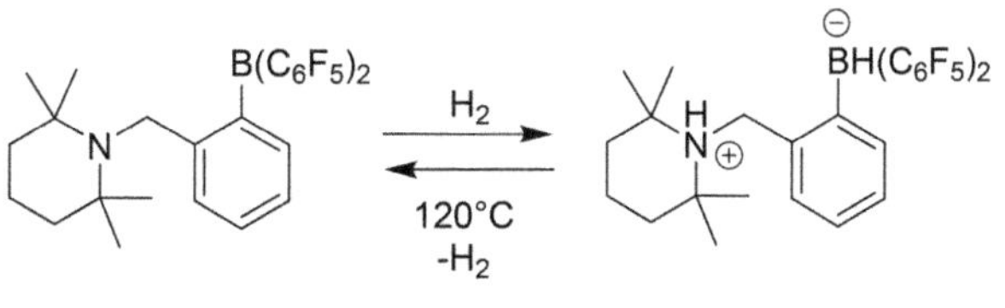

Scheme 2.21   Activation of dihydrogen by amine based FLPs.

binding has been studied *via* isotopic labeling and computations. These data support a transition state that is early in the activation of the dihydrogen molecule.[95]

In related work, Repo *et al.*[96,97] also re-investigated arene-linked N/B systems, initially described by Piers in 2003.[98] These authors modified the Piers system converting the NPh_2 fragment to dialkyl amines affording the intramolecular systems $C_6H_4(B(C_6F_5)_2)$ (NMe_2), $C_6H_4(B(C_6F_5)_2)(NC_5H_6Me_4)$, $C_6H_4(B(C_6F_5)_2)(CH_2NC_5H_6Me_4)$, and $C_6H_4(B(C_6F_5)_2)$ $(CH_2NC_9H_7Me_2i\text{-}Pr)$. The Repo group[99] also showed that while the related intramolecular FLP $C_6H_4(BH_2)(NC_5H_6Me_4)$ dimerizes, it does react with dihydrogen, affording the zwitterion $C_6H_4(BH_3)(NHC_5H_6Me_4)$ (Figure 2.5).

Erker and coworkers prepared the related N/B intramolecular FLPs, $C_5H_{10}NCH(Ph)$ $CH_2B(C_6F_5)_2$ and $C_5H_{10}NC_6H_{10}B(C_6F_5)_2$ (Figure 2.5)[100] by the hydroboration of the corresponding enamine with Piers' borane.[101,102] These species also activate dihydrogen in a facile manner. Fontaine and coworkers also reported that the related chloroborane $(C_6H_4NC_5H_6Me_4)_2BCl$ (Figure 2.5) activates dihydrogen to give the expected zwitterion,[103] Wang and coworkers[104] reported similar reactivity for N/B FLP derived from lutidine, $C_5H_3Me(CH_2B(C_6F_5)_2)N$. However, in the latter case, cleavage of the B–C bond affording $C_5H_3Me_2N$ and $HB(C_6F_5)_2$, or $[C_5H_8Me_2NH][H_2B(C_6F_5)_2]$ depending on the pressure of dihydrogen used (Figure 2.5).

More recently the Erker group[105] has developed a unique approach to the synthesis of an olefin linked N/B FLP. Reacting $C_5H_3Me_2NCH_2CCH$ and $(C_6F_5)_2BCH=CHR$ afforded the species $C_5H_3Me_2NCH_2C(CH_2CH=CHR)=CHB(C_6F_5)_2$. This species reacts with dihydrogen to give the zwitterion $C_5H_3Me_2NHCH_2C(CH_2CH=CHR)=C(H)(H)B(C_6F_5)_2$ (Scheme 2.22).

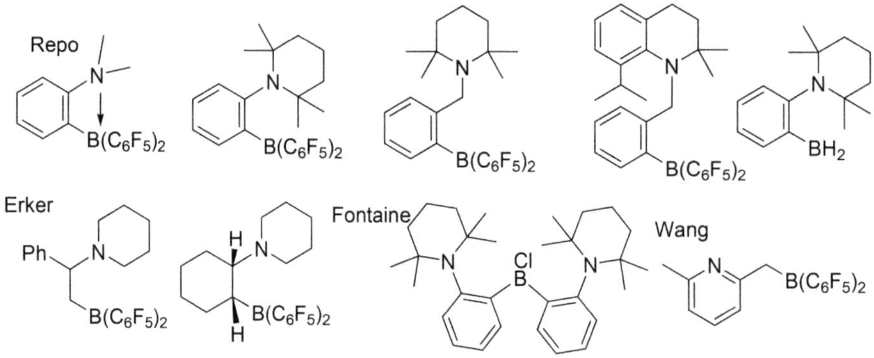

Figure 2.5   Intramolecular Amine/borane FLPs.

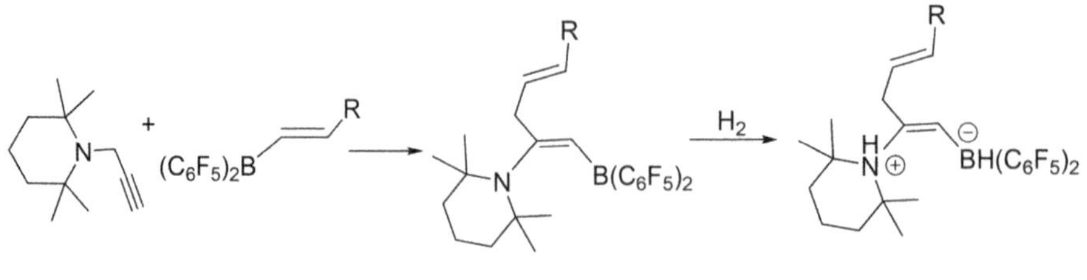

Scheme 2.22   Synthesis and reactivity of an olefin-linked N/B FLP.

2.10 FLPs Without Frustration

The above developments clearly show that frustrating the interaction of an electron donor with an electron acceptor provides a broad potential for FLP activation of dihydrogen. In a 2007 *Science* paper, Bertrand and coworkers[106] described the clean reaction of the alkylamino–carbene *i*-Pr$_2$NC*t*-Bu with dihydrogen, giving *i*-Pr$_2$NCH$_2$*t*-Bu (Scheme 2.23) at room temperature. This stands in contrast to the more common N-heterocyclic carbenes that do not react with dihydrogen. These observations are indeed reminiscent of the reaction of dihydrogen with FLPs. Carbene ligands are both electron donors and acceptors and, when suitably basic, the combined Lewis acidity and basicity result in the activation of dihydrogen. In contrast to FLPs, these characteristics reside on the same atom. In that sense, we could describe these species as 'FLPs without frustration' as the donor and acceptor sites are the same, albeit in orthogonal orbitals. This perspective is reinforced by recalling that Sander *et al.*[107–109] showed the carbene difluorovinylidene reacts with dihydrogen in an argon matrix at 20–30 K (Scheme 2.23). Moreover, this notion is further supported by the work of Power and coworkers[110,111] who showed that heavier congeners of carbenes, diaryldigermylenes, and diarylstannylenes also react with dihydrogen to give a mixture of products.

The requirement of steric frustration was probed by considering systems in which the Lewis acidic center and Lewis basic center are directly bound to each other. To this end, the monomeric species R$_2$PB(C$_6$F$_5$)$_2$ (R = Cy, *t*-Bu) were prepared from the reaction of secondary lithium phosphides (R$_2$PLi, R = Et, Ph, Cy, *t*-Bu) with (C$_6$F$_5$)$_2$BCl.[112] The sterically demanding nature of the substituents in these species precluded dimerization and also meant that phosphorus and boron centers retained their respective donor and acceptor properties. Indeed, these species undergo a slow reaction with dihydrogen (4 bar) at 60 °C to afford the phosphine–borane adducts (R$_2$PH)(BH(C$_6$F$_5$)$_2$) (R = Cy, *t*-Bu) in 48 h (Scheme 2.24). This dihydrogen activation was not reversible, as heating resulting only in the dissociation of the phosphine from the borane.

Crystallographic data for *t*-Bu$_2$PB(C$_6$F$_5$)$_2$ revealed a short B–P distance of 1.786(4) Å and pseudo-trigonal planar geometries about both the boron and phosphorus centers (Scheme 2.24). DFT calculations indicated that the π-bonding orbital between the boron and phosphorus atoms is the HOMO and it is significantly polarized. It is this polarization that presumably accounts for the reaction with dihydrogen. The boron–phosphorus distance in the product of (*t*-Bu$_2$PH)(BH(C$_6$F$_5$)$_2$) lengthens dramatically to 1.966(9) Å. Further DFT studies of this activation of dihydrogen infer an initial attack of dihydrogen at the Lewis acidic boron, while subsequent rotation presents the hydrogen–hydrogen bond parallel to the boron–phosphorus bond, allowing protonation of

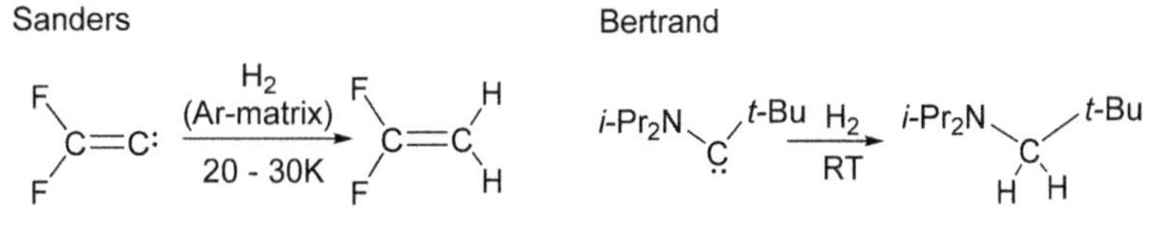

Scheme 2.23 Reduction of carbenes with dihydrogen.

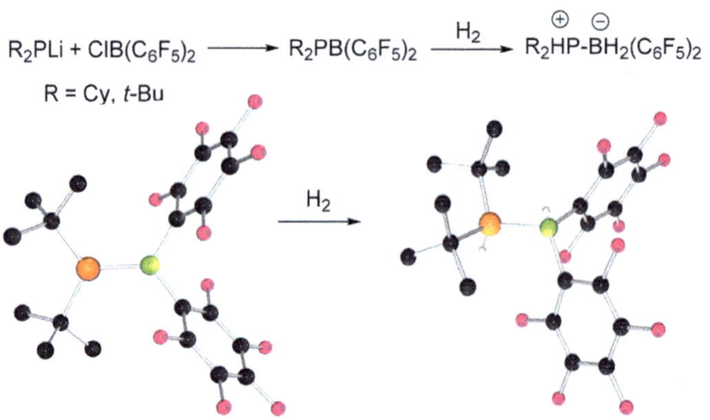

Scheme 2.24 Synthesis and reactions of phosphido–boranes with dihydrogen.

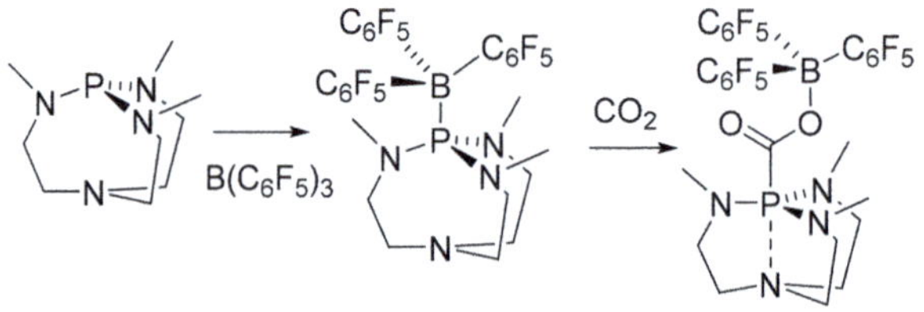

Scheme 2.25 FLP reactivity of classical Lewis acid–base adducts.

the phosphorus center. While the coordination of dihydrogen to boron has a barrier of *ca.* 22 kcal mol^{-1}, the remaining steps are barrierless. The overall reaction is exothermic (-43 kcal mol^{-1}) consistent with the irreversibility of the reaction.[112] In a more recent publication, Grubba and coworkers[113] reported the related reactions of the compound PhB(P*t*-Bu$_2$)$_2$ with dihydrogen affording the dimeric species [PhBH(HP*t*-Bu$_2$)]$_2$ and HP*t*-Bu$_2$. In contrast, the species *i*-Pr$_2$NB(PCy$_2$)$_2$ did not react with dihydrogen, presumably a result of the diminished Lewis acidity at boron.

The above chemistry indicates that FLP reactivity results from access to a lone pair of electrons and a vacant orbital and that steric frustration are *not* necessarily a requirement, but rather just one way for these features to be accessible. Indeed, we described the situation in R$_2$PB(C$_6$F$_5$)$_2$ species as 'electronic frustration', as there is an energy mismatch between the donor and acceptor orbitals on phosphorus and boron that inhibits π-bonding and makes these features susceptible to further reactivity.

A closely related but more recent example further illuminated an aspect of 'frustration'. The combination of the Verkade's superbase, N(CH$_2$CH$_2$NMe)$_3$P, with B(C$_6$F$_5$)$_3$ affords the robust Lewis acid–base adduct, N(CH$_2$CH$_2$NMe)$_3$PB(C$_6$F$_5$)$_3$,[114] which was both crystallographically characterized and shown to be stable in solution by NMR methods. Attempts to affect the exchange of the bound borane with additional equivalents of borane revealed no evidence of exchange. Nonetheless, this 'adduct' affects FLP chemistry with several small molecules including PhNCO, PhCH$_2$N$_3$, PhNSO, and CO$_2$ (Scheme 2.25). Preliminary work also showed that this species reacts with dihydrogen.

These observations demonstrate that even when a dissociative equilibrium is not observed on the NMR timescale, FLP reactivity may be observed, even in instances where 'frustration' is not apparent. Again, these results emphasize that classical Lewis acid–base adducts may be thermodynamically stable, but even a small dissociative equilibrium constant can lead to reactivity.

2.11 Implications

The above studies define several important aspects regarding the paradigm of FLP reactivity. Our initial understanding of the mechanism of activation of dihydrogen by the action of a Lewis acid and base was confined to the notion of the generation of an 'encounter-complex' and its reactivity with dihydrogen. More recently implications of the possibility of a radical mechanism have also been suggested. Such variations in the mechanism of reactions depend on the specific pairing of Lewis acid and base, but this broadens both the range of components for FLPs and their potential applications.

Variations of the nature and components of both intra- and intermolecular FLPs foreshadow a breadth of possible FLPs one could consider when optimizing reactivity for a specific application. Moreover, the scope of FLPs is further broadened by the systems that demonstrate that steric frustration is not essential. Indeed, systems that are electronically frustrated, or systems that can access equilibria involving Lewis acid–base adducts provide other systems capable of FLP reactivity.

With some knowledge of the mechanism, as well as several examples of systems capable of FLP reactivity, we noted that select systems could react with dihydrogen reversibly. These findings put us in a prime position to explore the potential of FLPs in hydrogenation catalysis. Indeed, this possibility was made even more exciting as we recognized very early on that such a process was unprecedented. The advent of metal-free hydrogenation catalysis is the subject of the next chapter.

References

1. B. S. Jursic, *J. Mol. Struct.*, 1999, **492**, 97–103.
2. J. D. Watts and R. J. Bartlett, *J. Am. Chem. Soc.*, 1995, **117**, 825–826.
3. T. J. J. Tague and L. Andrews, *J. Am. Chem. Soc.*, 1994, **116**, 4970–4976.
4. P. R. Schreiner, H. F. Schaefer and P. V. R. Schleyer, *J. Chem. Phys.*, 1994, **101**, 7625–7632.
5. A. Moroz and R. L. Sweany, *Inorg. Chem.*, 1992, **31**, 5236.
6. A. Moroz, R. L. Sweany and S. L. Whittenburg, *J. Phys. Chem.*, 1990, **94**, 1352.
7. T. A. Rokob, A. Hamza, A. Stirling, T. Soós and I. Papai, *Angew. Chem., Int. Ed.*, 2008, **47**, 2435–2438.
8. A. Stirling, A. Hamza, T. A. Rokob and I. Papai, *Chem. Commun.*, 2008, 3148–3150.
9. A. Hamza, A. Stirling, T. András Rokob and I. Pápai, *Int. J. Quantum Chem.*, 2009, **109**, 2416–2425.
10. T. Rokob, A. Hamza and I. Pápai, *J. Am. Chem. Soc.*, 2009, **131**, 10701–10710.
11. P. Spies, R. Fröhlich, G. Kehr, G. Erker and S. Grimme, *Chem. - Eur. J.*, 2008, **14**, 333–343.
12. T. A. Rokob, I. Bako, A. Stirling, A. Hamza and I. Papai, *J. Am. Chem. Soc.*, 2013, **135**, 4425–4437.
13. S. Grimme, H. Kruse, L. Goerigk and G. Erker, *Angew. Chem., Int. Ed.*, 2010, **49**, 1402–1405.
14. B. Schirmer and S. Grimme, *Chem. Commun.*, 2010, **46**, 7942–7944.
15. M. P. Pu and T. Privalov, *ChemPhysChem*, 2014, **15**, 3714–3719.
16. I. Bako, A. Stirling, S. Balint and I. Papai, *Dalton Trans.*, 2012, **41**, 9023–9025.
17. J. Daru, I. Bako, A. Stirling and I. Papai, *ACS Catal.*, 2019, **9**, 6049–6057.

18. M. Pu and T. Privalov, *J. Chem. Phys.*, 2013, **138**, 154305–154312.
19. M. P. Pu and T. Privalov, *ChemPhysChem*, 2014, **15**, 2936–2944.
20. D. J. Parks and W. E. Piers, *J. Am. Chem. Soc.*, 1996, **118**, 9440–9441.
21. S. Rendler and M. Oestreich, *Angew. Chem., Int. Ed.*, 2008, **47**, 5997–6000.
22. A. Y. Houghton, J. Hurmalainen, A. Mansikkamäki, W. E. Piers and H. M. Tuononen, *Nat. Chem.*, 2014, 983–988.
23. D. W. Stephan, *Nat. Chem.*, 2014, **6**, 952–953.
24. J. Chen and E. Y. X. Chen, *Angew. Chem.*, 2015, **127**, 6946–6950.
25. L. Rocchigiani, G. Ciancaleoni, C. Zuccaccia and A. Macchioni, *J. Am. Chem. Soc.*, 2014, **136**, 112–115.
26. H. Zaher, A. E. Ashley, M. Irwin, A. L. Thompson, M. J. Gutmann, T. Kramer and D. O'Hare, *Chem. Commun.*, 2013, **49**, 9755–9757.
27. L. C. Brown, J. M. Hogg, M. Gilmore, L. Moura, S. Imberti, S. Gartner, H. Q. N. Gunaratne, R. J. O'Donnell, N. Artioli, J. D. Holbrey and M. Swadzba-Kwasny, *Chem. Commun.*, 2018, **54**, 8689–8692.
28. A. Y. Houghton and T. Autrey, *J. Phys. Chem. A*, 2017, **121**, 8785–8790.
29. L. Koenczoel, E. Makkos, D. Bourissou and D. Szieberth, *Angew. Chem., Int. Ed.*, 2012, **51**, 9521–9524.
30. L. Liu, L. L. Cao, Y. Shao, G. Menard and D. W. Stephan, *Chem*, 2017, **3**, 259–267.
31. A. Merk, H. Grossekappenberg, M. Schmidtmann, M. P. Luecke, C. Lorent, M. Driess, M. Oestreich, H. F. T. Klare and T. Muller, *Angew. Chem., Int. Ed.*, 2018, **57**, 15267–15271.
32. F. Holtrop, A. R. Jupp, B. J. Kooij, N. P. Leest, B. Bruin and J. C. Slootweg, *Angew. Chem., Int. Ed.*, 2020, **59**, 22210–22216.
33. E. L. Bennett, E. J. Lawrence, R. J. Blagg, A. S. Mullen, F. Macmillan, A. W. Ehlers, D. J. Scott, J. S. Sapsford, A. E. Ashley, G. G. Wildgoose and J. C. Slootweg, *Angew. Chem., Int. Ed.*, 2019, **58**, 8362–8366.
34. H. D. Wang, R. Fröhlich, G. Kehr and G. Erker, *Chem. Commun.*, 2008, 5966–5968.
35. P. A. Chase, T. Jurca and D. W. Stephan, *Chem. Commun.*, 2008, 1701–1703.
36. V. Sumerin, F. Schulz, M. Nieger, M. Leskela, T. Repo and B. Rieger, *Angew. Chem., Int. Ed.*, 2008, **47**, 6001–6003.
37. F. Schulz, V. Sumerin, M. Leskelä, T. Repo and B. Rieger, *Dalton Trans.*, 2010, **39**, 1920–1922.
38. A. Ramos, A. J. Lough and D. W. Stephan, *Chem. Commun.*, 2009, 1118–1120.
39. D. P. Huber, G. Kehr, K. Bergander, R. Fröhlich, G. Erker, S. Tanino, Y. Ohki and K. Tatsumi, *Organometallics*, 2008, **27**, 5279–5284.
40. K. V. Axenov, G. Kehr, R. Fröhlich and G. Erker, *J. Am. Chem. Soc.*, 2009, **131**, 3454–3455.
41. K. V. Axenov, G. Kehr, R. Frohlich and G. Erker, *Organometallics*, 2009, **28**, 5148–5158.
42. E. Theuergarten, A. C. Tagne Kuate, M. Freytag and M. Tamm, *Isr. J. Chem.*, 2015, **55**, 202–205.
43. R. C. Neu, E. Y. Ouyang, S. J. Geier, X. Zhao, A. Ramos and D. W. Stephan, *Dalton Trans.*, 2010, **39**, 4285–4294.
44. M. Harhausen, R. Frohlich, G. Kehr and G. Erker, *Organometallics*, 2012, **31**, 2801–2809.
45. C. Jiang and D. W. Stephan, *Dalton Trans.*, 2013, **42**, 630–637.
46. M. Siedzielnik, K. Kaniewska-Laskowska, N. Szynkiewicz, J. Chojnacki and R. Grubba, *Polyhedron*, 2021, **194**, 114930.
47. G. E. Arnott, P. Moquist, C. G. Daniliuc, G. Kehr and G. Erker, *Eur. J. Inorg. Chem.*, 2014, **2014**, 1394–1398.
48. S. J. Geier and D. W. Stephan, *J. Am. Chem. Soc.*, 2009, **131**, 3476–3477.
49. S. J. Geier, A. L. Gille, T. M. Gilbert and D. W. Stephan, *Inorg. Chem.*, 2009, **48**, 10466–10474.
50. C. F. Jiang, O. Blacque, T. Fox and H. Berke, *Organometallics*, 2011, **30**, 2117–2124.
51. B. Ginovska, T. Autrey, K. Parab, M. E. Bowden, R. G. Potter and D. M. Camaioni, *Chem. - Eur. J.*, 2015, **21**, 15713–15719.
52. A. Karkamkar, K. Parab, D. M. Camaioni, D. Neiner, H. M. Cho, T. K. Nielsen and T. Autrey, *Dalton Trans.*, 2013, **42**, 615–619.
53. D. L. Wu, D. Z. Jia, L. Liu, L. Zhang and J. X. Guo, *J. Phys. Chem. A*, 2010, **114**, 11738–11745.
54. L. X. Dang, G. K. Schenter, T. M. Chang, S. M. Kathmann and T. Autrey, *J. Phys. Chem. Lett.*, 2012, 3, 3312–3319.
55. J. D. Webb, V. S. Laberge, S. J. Geier, D. W. Stephan and C. M. Crudden, *Chem. - Eur. J.*, 2010, **16**, 4895–4902.
56. C. B. Caputo, K. L. Zhu, V. N. Vukotic, S. J. Loeb and D. W. Stephan, *Angew. Chem., Int. Ed.*, 2013, **52**, 960–963.
57. S. Bhunya and A. Paul, *Chem. - Eur. J.*, 2013, **19**, 11541–11546.
58. M. H. Holthausen, M. Colussi and D. W. Stephan, *Chem. - Eur. J.*, 2015, **21**, 2193–2199.
59. P. A. Chase and D. W. Stephan, *Angew. Chem., Int. Ed.*, 2008, **47**, 7433–7437.

60. P. A. Chase, A. L. Gille, T. M. Gilbert and D. W. Stephan, *Dalton Trans.*, 2009, 7179–7188.
61. D. Holschumacher, T. Bannenberg, C. G. Hrib, P. G. Jones and M. Tamm, *Angew. Chem., Int. Ed.*, 2008, **47**, 7428–7432.
62. D. Holschumacher, C. Taouss, T. Bannenberg, C. G. Hrib, C. G. Daniliuc, P. G. Jones and M. Tamm, *Dalton Trans.*, 2009, 6927–6929.
63. S. Kronig, E. Theuergarten, D. Holschumacher, T. Bannenberg, C. G. Daniliuc, P. G. Jones and M. Tamm, *Inorg. Chem.*, 2011, **50**, 7344–7359.
64. E. L. Kolychev, T. Bannenberg, M. Freytag, C. G. Daniliuc, P. G. Jones and M. Tamm, *Chem. - Eur. J.*, 2012, **18**, 16938–16946.
65. Y. Hoshimoto, T. Kinoshita, M. Ohashi and S. Ogoshi, *Angew. Chem., Int. Ed.*, 2015, 11666–11671.
66. G. C. Welch and D. W. Stephan, *J. Am. Chem. Soc.*, 2007, **129**, 1880–1881.
67. M. Ullrich, A. J. Lough and D. W. Stephan, *J. Am. Chem. Soc.*, 2009, **131**, 52–53.
68. M. Ullrich, A. J. Lough and D. W. Stephan, *Organometallics*, 2010, **29**, 3647–3654.
69. M. Pu, M. Heshmat and T. Privalov, *J. Chem. Phys.*, 2017, **147**, 014303/014301–014303/014315.
70. K. Takeuchi and D. W. Stephan, *Chem. Commun.*, 2012, **48**, 11304–11306.
71. C. F. Jiang, O. Blacque, T. Fox and H. Berke, *Dalton Trans.*, 2011, **40**, 1091–1097.
72. S. C. Binding, H. Zaher, F. M. Chadwick and D. O'Hare, *Dalton Trans.*, 2012, **41**, 9061–9066.
73. T. J. Herrington, A. J. W. Thom, A. J. P. White and A. E. Ashley, *Dalton Trans.*, 2012, **41**, 9019–9022.
74. Z. P. Lu, Z. H. Cheng, Z. X. Chen, L. H. Weng, Z. H. Li and H. D. Wang, *Angew. Chem., Int. Ed.*, 2011, **50**, 12227–12231.
75. R. J. Blagg, E. J. Lawrence, K. Resner, V. S. Oganesyan, T. J. Herrington, A. E. Ashley and G. G. Wildgoose, *Dalton Trans.*, 2016, **45**, 6023–6031.
76. R. J. Blagg, T. R. Simmons, G. R. Hatton, J. M. Courtney, E. L. Bennett, E. J. Lawrence and G. G. Wildgoose, *Dalton Trans.*, 2016, **45**, 6032–6043.
77. R. J. Blagg and G. G. Wildgoose, *RSC Adv.*, 2016, **6**, 42421–42427.
78. P. J. Hill, T. J. Herrington, N. H. Rees, A. J. P. White and A. E. Ashley, *Dalton Trans.*, 2015, **44**, 8984–8992.
79. P. Spies, S. Schwendemann, S. Lange, G. Kehr, R. Fröhlich and G. Erker, *Angew. Chem., Int. Ed.*, 2008, **47**, 7543–7546.
80. L. Liu Zeonjuk, P. St. Petkov, T. Heine, G.-V. Roeschenthaler, J. Eicher and N. Vankova, *Phys. Chem. Chem. Phys.*, 2015, **17**, 10687–10698.
81. T. Oezguen, K.-Y. Ye, C. G. Daniliuc, B. Wibbeling, L. Liu, S. Grimme, G. Kehr and G. Erker, *Chem. - Eur. J.*, 2016, **22**, 5988–5995.
82. M. Sajid, G. Kehr, T. Wiegand, H. Eckert, C. Schwickert, R. Pottgen, A. J. P. Cardenas, T. H. Warren, R. Fröhlich, C. G. Daniliuc and G. Erker, *J. Am. Chem. Soc.*, 2013, **135**, 8882–8895.
83. A. Ueno, X. Tao, C. G. Daniliuc, G. Kehr and G. Erker, *Organometallics*, 2018, **37**, 2665–2668.
84. K. Watanabe, A. Ueno, X. Tao, K. Skoch, X. Jie, S. Vagin, B. Rieger, C. G. Daniliuc, M. C. Letzel, G. Kehr and G. Erker, *Chem. Sci.*, 2020, **11**, 7349–7355.
85. C. Chen, C. G. Daniliuc, C. Mueck-Lichtenfeld, G. Kehr and G. Erker, *Chem. Commun.*, 2020, **56**, 8806–8809.
86. L. Wang, S. Dong, C. G. Daniliuc, L. Liu, S. Grimme, R. Knitsch, H. Eckert, M. R. Hansen, G. Kehr and G. Erker, *Chem. Sci.*, 2018, **9**, 1544–1550.
87. X. Jie, C. G. Daniliuc, R. Knitsch, M. R. Hansen, H. Eckert, S. Ehlert, S. Grimme, G. Kehr and G. Erker, *Angew. Chem., Int. Ed.*, 2019, **58**, 882–886.
88. T. Özgün, G.-Q. Chen, C. G. Daniliuc, A. C. McQuilken, T. H. Warren, R. Knitsch, H. Eckert, G. Kehr and G. Erker, *Organometallics*, 2016, **35**, 3667–3680.
89. G. Q. Chen, G. Kehr, C. G. Daniliuc, C. Mück-Lichtenfeld and G. Erker, *Angew. Chem.*, 2016, **128**, 5616–5620.
90. R. Knitsch, T. Oezguen, G.-Q. Chen, G. Kehr, G. Erker, M. R. Hansen and H. Eckert, *ChemPhysChem*, 2019, **20**, 1837–1849.
91. S. Dong, C. G. Daniliuc, G. Kehr and G. Erker, *Chem. - Eur. J.*, 2020, **26**, 745–753.
92. Z. Mo, E. L. Kolychev, A. Rit, J. Campos, H. Niu and S. Aldridge, *J. Am. Chem. Soc.*, 2015, **137**, 12227–12230.
93. F. Kutter, E. Lork and J. Beckmann, *Z. Anorg. Allg. Chem.*, 2018, **644**, 1234–1237.
94. V. Sumerin, F. Schulz, M. Atsumi, C. Wang, M. Nieger, M. Leskelä, T. Repo, P. Pyykkö and B. Rieger, *J. Am. Chem. Soc.*, 2008, **130**, 14117–14118.
95. F. Schulz, V. Sumerin, S. Heikkinen, B. Pedersen, C. Wang, M. Atsumi, M. Leskelä, T. Repo, P. Pyykkö, W. Petry and B. Rieger, *J. Am. Chem. Soc.*, 2011, **133**, 20245–20257.
96. V. Sumerin, K. Chernichenko, M. Nieger, M. Leskelä, B. Rieger and T. Repo, *Adv. Synth. Catal.*, 2011, **353**, 2093–2110.

97. K. Chernichenko, M. Nieger, M. Leskelä and T. Repo, *Dalton Trans.*, 2012, **41**, 9029–9032.

98. R. Roesler, W. E. Piers and M. Parvez, *J. Organomet. Chem.*, 2003, **680**, 218–222.

99. K. Chernichenko, B. Kotai, I. Papai, V. Zhivonitko, M. Nieger, M. Leskela and T. Repo, *Angew. Chem., Int. Ed.*, 2015, **54**, 1749–1753.

100. S. Schwendemann, R. Fröhlich, G. Kehr and G. Erker, *Chem. Sci.*, 2011, **2**, 1842–1849.

101. D. J. Parks, R. E. V. H. Spence and W. E. Piers, *Angew. Chem., Int. Ed.*, 1995, **34**, 809–811.

102. D. J. Parks, W. E. Piers and G. P. A. Yap, *Organometallics*, 1998, **17**, 5492–5503.

103. M.-A. Courtemanche, E. Rochette, M.-A. Legare, W. Bi and F.-G. Fontaine, *Dalton Trans.*, 2016, **45**, 6129–6135.

104. J. Zheng, Y.-J. Lin and H. Wang, *Dalton Trans.*, 2016, **45**, 6088–6093.

105. T. Wang, C. G. Daniliuc, C. Muck-Lichtenfeld, G. Kehr and G. Erker, *J. Am. Chem. Soc.*, 2018, **140**, 3635–3643.

106. G. D. Frey, V. Lavallo, B. Donnadieu, W. W. Schoeller and G. Bertrand, *Science*, 2007, **316**, 439–441.

107. C. Kötting and W. Sander, *J. Am. Chem. Soc.*, 1999, **121**, 8891–8897.

108. W. Sander and C. Koetting, *Chem. - Eur. J.*, 1999, **5**, 24–28.

109. W. Sander and C. Kötting, *Chem. - Eur. J.*, 1999, **5**, 24–28.

110. Y. Peng, B. D. Ellis, X. Wang and P. P. Power, *J. Am. Chem. Soc.*, 2008, **130**, 12268–12269.

111. G. H. Spikes, J. C. Fettinger and P. P. Power, *J. Am. Chem. Soc.*, 2005, **127**, 12232–12233.

112. S. J. Geier, T. M. Gilbert and D. W. Stephan, *J. Am. Chem. Soc.*, 2008, **130**, 12632–12633.

113. N. Szynkiewicz, A. Ordyszewska, J. Chojnacki and R. Grubba, *Inorg. Chem.*, 2021, **60**, 3794–3806.

114. T. C. Johnstone, G. N. J. H. Wee and D. W. Stephan, *Angew. Chem., Int. Ed.*, 2018, **57**, 5881–5884.

3 Borane-based FLP Hydrogenations

3.1 Chapter Overview

The ability of FLPs to activate dihydrogen reversibly begs the question of their potential to deliver proton and hydride to a substrate molecule. If this can be achieved, then indeed the FLP is available for further activation of dihydrogen and thus becomes a catalyst for hydrogenation. In this chapter, we discuss the early results that demonstrate that such metal-free reductions are indeed possible. We begin by setting the stage as it was in the early 2000s, describing the limited literature, and then describing the early findings establishing FLP hydrogenations. Subsequent studies that broadened the substrate scope, extended the limits of functional group tolerance, and facilitated ease of handling are then considered. In this chapter, we limit discussion to FLP systems invoking the use of borane-based Lewis acids as these systems dominated the early developments.

3.2 Literature Precedents for Metal-free Hydrogenations

It is important to recognize that at the time that FLP hydrogenations were conceived, the concept of metal-free hydrogenations was not without precedent. In 1961, Walling *et al.*[1,2] had observed the reduction of benzophenone to diphenylmethanol with dihydrogen in the presence of potassium *t*-butoxide. However, the reaction conditions were quite forcing, as these reductions required dihydrogen pressures of over 100 bar and temperatures of *ca.* 200 °C.[3] In 2002, Berkessel *et al.* subsequently studied this reaction[4] and proposed a reaction pathway (Scheme 3.1) which they likened to that seen for the Ru-based Noyori hydrogenation of ketones.[5,6]

Metal-free hydrogenation was also known to be mediated by strong acids under forcing conditions. For example, aromatic hydrocarbons, cyclic alkenes and dienes were hydrogenated using dihydrogen at elevated pressure, in the presence of strong acids

A Primer in Frustrated Lewis Pair Hydrogenation: Concepts to Applications
By Douglas W. Stephan
© Douglas W. Stephan 2022
Published by the Royal Society of Chemistry, www.rsc.org

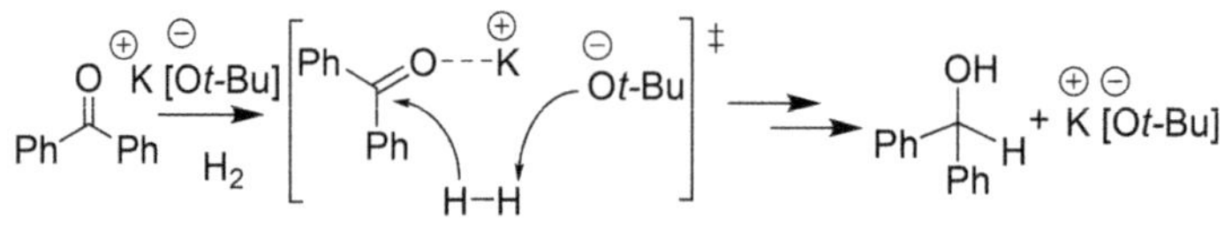

Scheme 3.1 Reduction of benzophenone by KOt-Bu/dihydrogen.

such as HF–TaF$_5$, HF–SbF$_5$, or HBr–AlBr$_3$, affording the saturated products and rearrangement of the carbon frameworks in some cases.[7–9] Borane-catalyzed hydrogenation of arenes had been studied by Köster and coworkers.[10,11] These authors showed that fully or partly hydrogenated derivatives were obtained, although again, high temperatures (~200 °C) and high dihydrogen pressures were required. Similarly, Haenel described a related procedure using a homogeneous borane catalyst for the liquefaction of coal.[12] Despite these precedents, catalytic hydrogenation based on transition metal catalysts had been widely exploited, and indeed the belief that metals were necessary for this process was accepted as a requisite for effective hydrogenation under mild conditions.

3.3 FLP-hydrogenation Catalysis: the Early Findings

The reversible activation of dihydrogen by FLPs prompted consideration of the possibility of metal-free hydrogenation catalysis. We envisioned sequential delivery of proton and hydride to the substrate regenerating the FLP providing a cyclic process for the activation of dihydrogen. To achieve catalysis, it was also recognized that neither the substrate nor the product should form an adduct with the Lewis acid, as this would sequester the Lewis acid and preclude catalysis. Thus, the first attempts targeted sterically encumbering imine substrates (Scheme 3.2).[13,14] Stoichiometric addition of an imine to [Mes$_2$PH(C$_6$F$_4$)BH(C$_6$F$_5$)$_2$] at room temperature provided the borane-amine adduct [Mes$_2$P(C$_6$F$_4$)B(C$_6$F$_5$)$_2$(NHRCH$_2$R′)], resulting from the transfer of both proton and hydride to the imine. Thermally inducing amine dissociation regenerated the phosphine–borane Mes$_2$P(C$_6$F$_4$)B(C$_6$F$_5$)$_2$ allowing it to activate a further equivalent of dihydrogen reforming [Mes$_2$PH(C$_6$F$_4$)BH(C$_6$F$_5$)$_2$] and thus initiating a catalytic cycle. In this fashion, the electron-rich imine, t-BuN=CPh(H) was catalytically reduced to the corresponding amine in the presence of 5 mol% of the phosphonium-hydridoborate, [Mes$_2$PH(C$_6$F$_4$)BH(C$_6$F$_5$)$_2$] under an atmosphere of dihydrogen (1.5 bar) at 80 °C. Similarly, the electron-poor imine, PhSO$_2$N=CPh(H) was reduced, albeit significantly more slowly even on heating to 120 °C.

Analogous catalytic reductions of other imines proceeded when bulky substituents on the nitrogen atom were present, affording the corresponding amines in high isolated yields.[13] Imines with less sterically demanding substituents such as a benzyl group were only stoichiometrically reduced. This was attributed to the irreversible coordination of the amine to the boron center precluding further activation of dihydrogen. Analogous treatment of the N-aryl aziridine (PhCH)$_2$NPh also afforded hydrogenation to the ring-opened amine, PhCHCH(Ph)NHPh. These catalytic reductions proved to be 'living' in as much as the addition of more substrate to reaction mixtures prompted further catalysis.

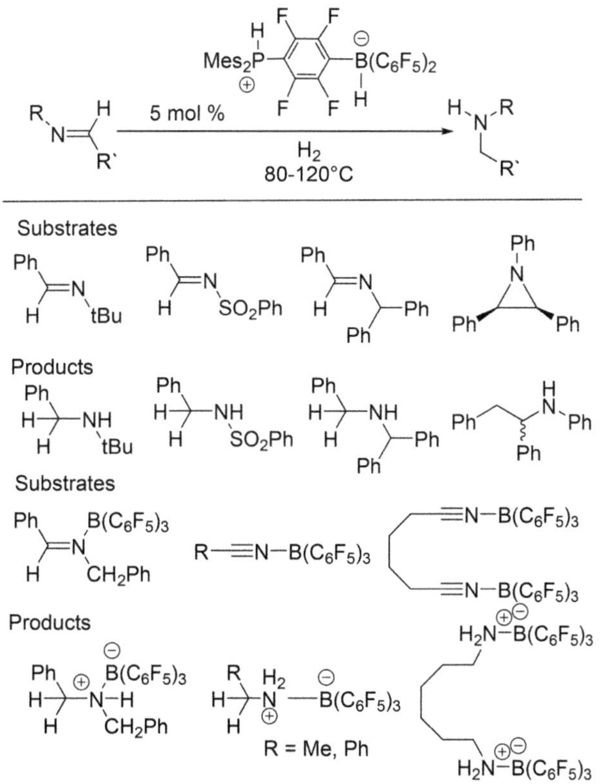

Scheme 3.2 Metal-free catalytic hydrogenation of imines, aziridines, and protected nitriles.

In the case of sterically less encumbered imines and nitriles, coordination to $B(C_6F_5)_3$ provided an effective protecting group to preclude interaction with the boron atom of the catalyst. Thus, the carbon–nitrogen triple bonds of borane–nitrile adducts could be hydrogenated, affording the amine-$B(C_6F_5)_3$ adducts as products, albeit at somewhat slower rates of conversion. This protocol relies on the greater Lewis acidity of $B(C_6F_5)_3$ over that of the boron center in the phosphine–borane catalyst. In this fashion, catalyst inhibition by the substrate or the reduced products can be avoided. While the latter procedure extends metal-free reductions nitriles, it should be noted that this was done simply as a demonstration of principle rather than a cost-effective protocol.

The mechanism for this catalytic cycle was proposed to involve initial protonation of the substrate followed by hydride delivery (Scheme 3.3). This view was supported by the diminished reactivity of less basic imines, suggesting that the rate-determining step in the reduction involves protonation of the imine. In addition, the use of the phosphonium-borate $(Cy_3P)(C_6F_4)BH(C_6F_5)_2$[15] as the catalyst showed no reaction. Lastly, the reaction of $[Mes_2PH(C_6F_4)BH(C_6F_5)_2]$ and the imine $PhC(t\text{-}Bu)=N(t\text{-}Bu)$ gave the salt $[PhC(t\text{-}Bu)=NH(t\text{-}Bu)][Mes_2P(C_6F_4)BH(C_6F_5)_2]$. In the latter case, presumably, while the basicity of the imine prompts protonation, the steric congestion precludes hydride delivery to the ketimine carbon.[13]

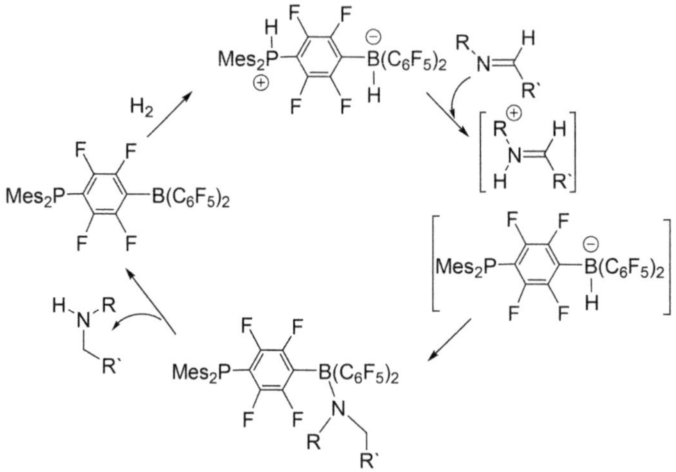

Scheme 3.3 Proposed mechanism for metal-free catalytic hydrogenations.

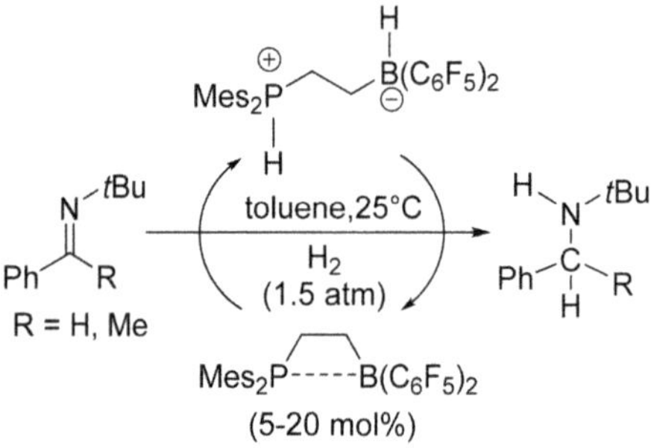

Scheme 3.4 Catalytic hydrogenation of imines using $Mes_2PC_2H_4B(C_6F_5)_2$.

Shortly after the initial report of FLP hydrogenation, Erker and coworkers[16] described the use of the intramolecular FLP, $Mes_2PC_2H_4B(C_6F_5)_2$ in the reduction of imines. This species was found to be more active than that derived from the zwitterion $[Mes_2PH(C_6F_4)BH(C_6F_5)_2]$, as Erker's FLP mediated reductions under ambient conditions. For example, the aldimine PhCH=Nt-Bu was reduced at 25 °C under dihydrogen (1.5 bar) using 20 mol% $Mes_2PC_2H_4B(C_6F_5)_2$ (Scheme 3.4). This catalyst was also effective in the reduction of the ketimine PhCMe=Nt-Bu, using only 5 mol% of the intramolecular FLP catalyst.

The Erker group also reasoned that if imine reduction proceeds *via* an iminium intermediate, then reduction of enamines should also be viable.[16] Indeed, this is the case. Using 10 mol% of $Mes_2PC_2H_4B(C_6F_5)_2$, the reduction of the enamine $PhC(NC_5H_{10})=CH_2$ was shown to give the amine $PhCH(NC_5H_{10})Me$ under mild conditions (25 °C, 1.5 bar). In this fashion, several related enamines were reduced using as little as 3 mol% catalyst (Scheme 3.5).[16] The enamine $PhC(NC_4H_8O)=CH_2$ required the use of more forcing conditions (50 bar, 80 °C) to give the corresponding amine product in 80% yield.[16] In subsequent studies, details of the kinetics and thermodynamics[17] of imine hydrogenation have been studied by a combination of time-resolved reaction

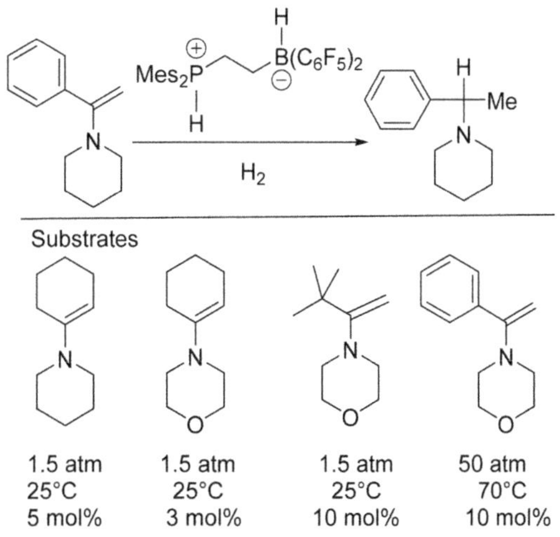

Scheme 3.5 Catalytic hydrogenation of enamines.

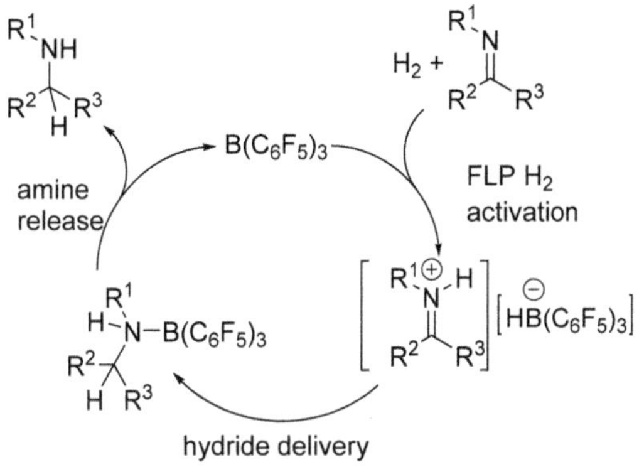

Scheme 3.6 Proposed mechanism of catalytic hydrogenation of imines by $B(C_6F_5)_3$.

calorimetry and NMR spectroscopy using $Mes_2PCH_2CH_2B(C_6F_5)_2$ as the catalyst. These experiments revealed a turnover frequency (TOF) of 1.1 min^{-1} with an enthalpic driving force of *ca.* −73 kJ mol^{-1}. In addition, the FLP, $Mes_2PCH_2CH_2B(C_6F_5)_2$ exhibited a rate constant for the heterolytic splitting of dihydrogen of $k = 0.7(3)$ M^{-1} s^{-1} with a ΔG of −29.8 kJ mol^{-1}.

With the initial report describing the reduction of bulky imines, we recognized that it should be possible to simplify FLP catalysts as one could exploit the substrate as the basic component of the FLP.[14] To test this notion, a catalytic amount of $B(C_6F_5)_3$ was combined with a bulky imine substrate under dihydrogen. This led to the catalytic reduction of the imine to the corresponding amine. Indeed, this strategy was effective for several imines. In this simplified FLP reduction, the imine substrate, and the Lewis acid act as an FLP to activate dihydrogen generating an iminium cation and hydridoborate. Subsequent delivery of the hydride to the iminium carbon yields the product amine freeing the borane for further reaction (Scheme 3.6).[14] In the cases where phosphine was required, the mechanism is similar to that described above for $Mes_2P(C_6F_4)B(C_6F_5)_2$ where the transient phosphonium cation transfers proton to the nitrogen of the imine prompting hydride transfer and thus reduction.

Interestingly, for poorly basic imines, the addition of a catalytic amount of the phosphine $PMes_3$ accelerated the hydrogenation (8 h *vs.* 41 h), as the reaction of phosphine/borane with dihydrogen proceeded more rapidly than that with the weakly basic imine. A similar strategy using a catalytic amount of $PMes_3$ and $B(C_6F_5)_3$ under dihydrogen was also employed to reduce nitrile-borane adducts. It is noteworthy that in the absence of phosphine, no reduction of the nitrile-borane adducts was observed (Scheme 3.7).[14]

Almost concurrently with the work above, Chen and Klankermayer[18] reported analogous reductions employing $B(C_6F_5)_3$ alone in combination with imine substrates. In an important precedent-setting experiment, these authors described the first efforts to affect the asymmetric reduction of PhN=CPh(Me) using the chiral borane (α-pinenyl) $B(C_6F_5)_2$. While this resulted in the modest 13% enantiomeric excess in the product amine, this result foreshadowed extensive efforts that have dramatically improved enantioselectivity (*vide* Chapter 4).[18]

In a further extension, Erker and coworkers[19] showed that a combination of 20 mol% of the Lewis base $C_{10}H_6(PPh_2)_2$ and the Lewis acid $B(C_6F_5)_3$ affects the hydrogenation of a variety of silyl enol ethers at room temperature (Scheme 3.8). In the case of the silyl enol-ether $Me_3SiOCMe=CH_2$, only stoichiometric hydrogenation was observed under these mild conditions; however, at higher dihydrogen pressure and temperature (60 bar H_2, 70 °C), catalytic reduction was achieved.

The Berke group[20] employed 10 mol% of the Lewis acid 1,8-bis(pentafluorophenylboryl) naphthalene, $C_{10}H_6(B(C_6F_5)_2)_2$ under 15 bar dihydrogen at 120 °C to affect the hydrogenation of imines (Scheme 3.9). Interestingly, these authors suggested that in this case, dihydrogen is not activated between the two boron centers as this pathway was computed to have a higher barrier than the activation of dihydrogen at a single boron center.

Repo and Rieger[21] demonstrated that the intramolecular nitrogen–boron zwitterion $[C_5H_6Me_4NCH_2C_6H_4BH(C_6F_5)_2]$ could also mediate the catalytic hydrogenation of imines and enamines. (Scheme 3.10).[21] Using 4 mol% of this catalyst gave near

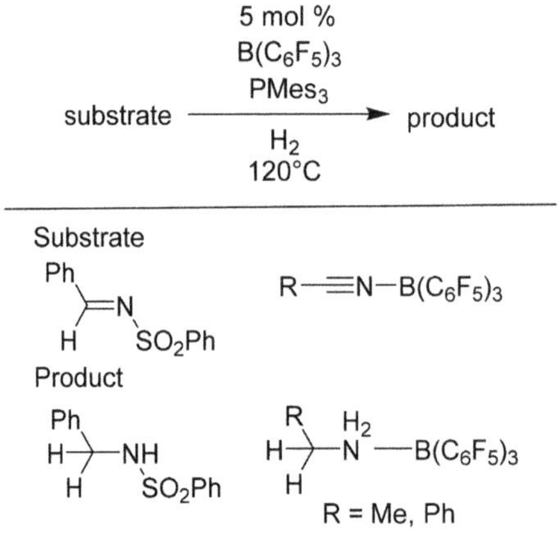

Scheme 3.7 Catalytic hydrogenation of selected imines by $B(C_6F_5)_3$ in presence of $PMes_3$.

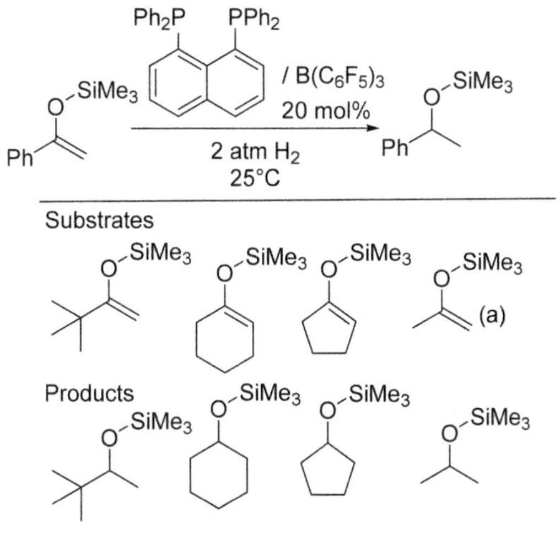

Scheme 3.8 Catalytic hydrogenation of silyl-enol ethers; (a) 60 bar H₂, 70 °C.

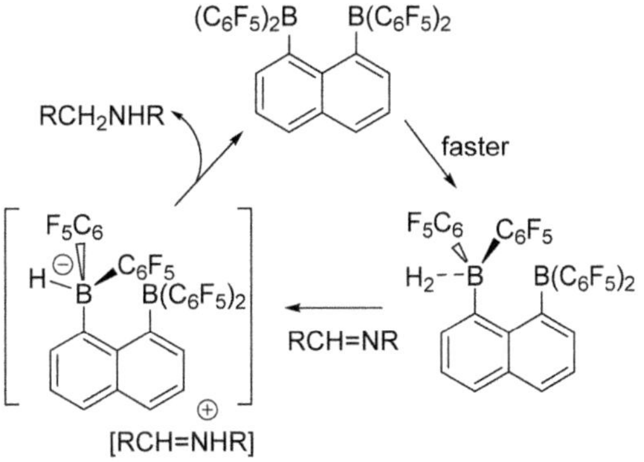

Scheme 3.9 Catalytic of H₂ by bis-borane FLP catalyst.

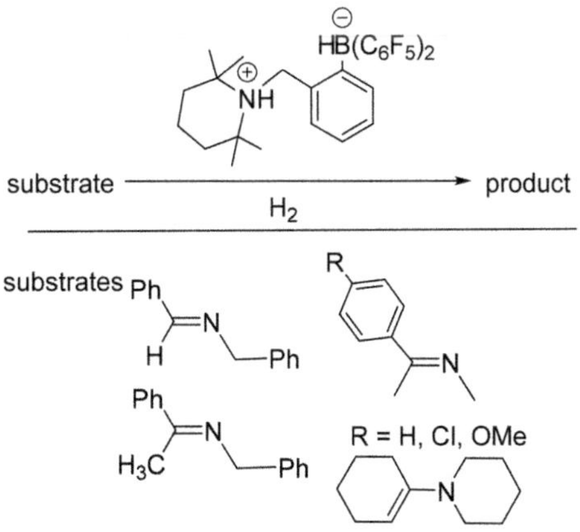

Scheme 3.10 Catalytic hydrogenation of selected imines and enamines by [C₅H₆Me₄NHCH₂C₆H₄BH(C₆F₅)₂].

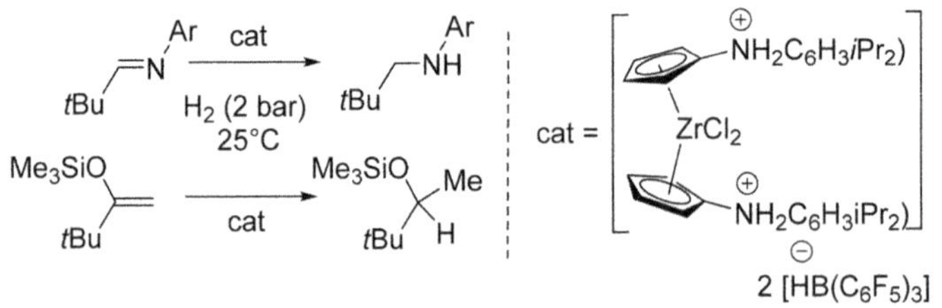

Scheme 3.11 Hydrogenation using the Zr-salt, $[(C_5H_4CH_2NH_2(C_6H_3i\text{-}Pr_2))_2ZrCl_2][HB(C_6F_5)_3]_2$.

quantitative reductions of sterically encumbered substrates, whereas those that were less encumbered such as $PhCH_2C(Me)=NMe$ and $RC_6H_4C(Me)=NMe$ (R = H, Cl, OMe) were only stoichiometrically reduced.

In further extending the nature of FLP catalysts, the Erker group[22] demonstrated that the combination of the basic diimine-species $[(C_5H_4CH=N(C_6H_3i\text{-}Pr_2))_2ZrCl_2]$ in the presence of Lewis acids $B(C_6F_5)_3$ affected the hydrogenation of the imine in the Zr-complex to give the bis-ammonium zirconocene-salt, $[(C_5H_4CH_2NH_2(C_6H_3i\text{-}Pr_2))_2ZrCl_2]$ $[HB(C_6F_5)_3]_2$. Moreover, this product was also capable of mediating the FLP hydrogenation of a sterically encumbered imine (Scheme 3.11) as well as a silyl enol ether, thus extending FLP catalysts to include systems incorporating ancillary metal centers.

The above findings, primarily reported between 2007 and early 2009, confirmed the notion that the activation of dihydrogen by FLPs can be exploited to deliver proton and hydride to judiciously selected substrates catalytically. While these data establish the viability of metal-free hydrogenation catalysis, they also prompt several questions concerning the substrate scope, the nature of the optimal conditions for catalysis, the impact of FLP catalyst variation, and functional group tolerance. The following Sections 3.4–3.6 address these issues.

3.4 Borane Variation and Mechanism in Imine Reduction

Insight into the impact of variation of the Lewis acid component of the FLP was first probed by Paradies *et al.*[23] These researchers reported a detailed kinetic study employing the fluorinated boranes $B(C_6F_5)_3$, $B(2,4,6\text{-}C_6F_3H_2)_3$ or $B(2,6\text{-}C_6F_2H_3)_3$ in imine hydrogenation. The study of these systems revealed divergent mechanisms for hydrogenation with these boranes. The kinetics confirmed that $B(C_6F_5)_3$ operates in concert with the imine to activate dihydrogen. On the other hand, the less electrophilic Lewis acids affect imine reduction *via* a mechanism in which the amine product and the borane act on dihydrogen (Scheme 3.12). This auto-induced mechanism involving activation of dihydrogen by amine and borane has a ΔG that is 2 kcal mol^{-1} lower than when imine participation is considered. Interestingly, the deuteration of imine showed an unusual inverse isotope effect. This was attributed to the higher exothermicity of D_2-activation. In a subsequent paper, Paradies[24] exploited $B(2,6\text{-}C_6F_2H_3)_3$ as the catalyst to affect the reduction of 16 imines. The auto-induced catalysis gave rate constants that were 8–10 times higher than those seen for the simple catalytic cycle. In addition, the simple catalytic cycle was found to be more susceptible to electronic perturbations that altered the pK_a of the imine.

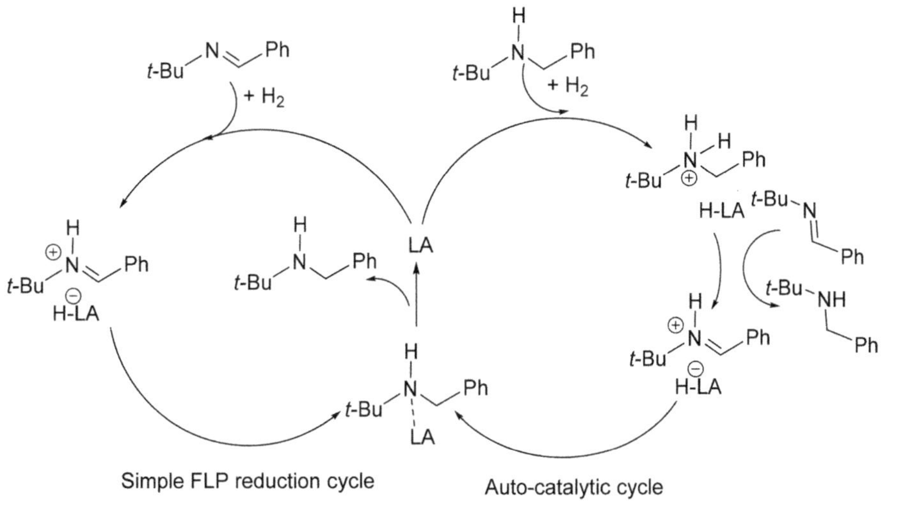

Scheme 3.12 Simple and auto-induced catalytic cycles for FLP imine reduction.

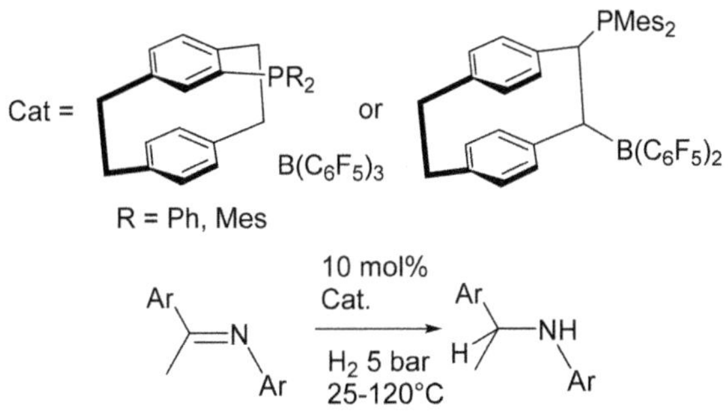

Scheme 3.13 Imine hydrogenation using [2.2]paracyclophane-derived FLPs.

In another study using these optimized boranes, Paradies and co-workers[25] probed reductions where P/B FLPs were used for imine reductions. They demonstrated that there is also a strong dependence of the rate of hydrogenation on the electronic nature of the phosphine and the acidity of the corresponding phosphonium cation. These findings inferred that a careful balance of the acidity and basicity of the FLP components are required for highly efficient metal-free hydrogenation catalysts. In yet another paper, Paradies' group demonstrated a further strategy for the acceleration of the hydrogenation of imines.[26] Using microwave heating, $B(C_6F_5)_3$ mediated reductions were accelerated by a factor of 2.5 under dihydrogen (4 bar). This methodology also was applied to the reduction of enamines and N-based heterocycles.

Imine reductions have also been reported employing a series of [2.2]paracyclophane-derived FLPs.[27] Thus 10 mol% of the intermolecular FLP derived from $C_6H_3(PR_2)$ $(CH_2CH_2)_2C_6H_4$ and $B(C_6F_5)_3$ or the intramolecular FLP $(C_6H_4)_2(CH_2CH_2)(CHPR_2)$ $CHB(C_6F_5)_2)$ was used to hydrogenate a series of imines in generally good yield at temperatures ranging from 25–120 °C and 5 bar dihydrogen pressure over 6–30 h (Scheme 3.13). Similarly, intermolecular FLPs derived from paracyclophane-phosphines and $B(C_6F_5)_3$ have been used for silyl enol ether reductions.[28]

3.5   Functional Group Tolerance in Imine Reduction

A facile approach to assess the functional group tolerance for FLP-mediated imine hydrogenations was employed rather than preparing a large library of imine substrates. Functionalized molecular additives at concentrations equal to that of the readily available substrate, PhCH=Nt-Bu were employed to test the catalyst tolerance for functional groups. In the presence of naphthalene, bulky ethers, n-hexyl acrylate, bulky amines or alkyl or aryl halides,[29] the catalysts, B(C$_6$F$_5$)$_3$ or Mes$_2$PC$_6$F$_4$B(C$_6$F$_5$)$_2$ effectively reduced the imine substrate. However, additives such as PhNMe$_2$, t-BuNH$_2$, carbamate esters, ketones or aldehydes stifled catalysis. Similarly, these catalysts were ineffective in the presence of MesOH but tolerated 2,6-t-Bu$_2$C$_6$H$_3$OH. These data show that the first generation of FLP catalysts only tolerates functional groups that are either non-polar or sterically encumbered.

Targeting enhanced functional group tolerance, the Soós group[30] employed a strategy that they described as 'size exclusion'. This concept required precise control of the environment around the boron center. To this end, these researchers prepared the sterically congested Lewis acid center B(C$_6$F$_5$)$_2$Mes (Figure 3.1) anticipating that this species would be too congested to interact with donor functional groups, while still being accessible for reaction with dihydrogen. Indeed, combining this Lewis acid with one of the nitrogen bases, CH(CH$_2$CH$_2$)$_3$N or N(CH$_2$CH$_2$)$_3$N, provided an effective catalyst for the reduction of imines at 20 °C and 4 bar of dihydrogen,[30] thus validating this catalyst design concept.

Ashley and co-workers[31] used a similar strategy, developing the air-stable Lewis acids B(C$_6$Cl$_5$)$_2$(C$_6$F$_5$) and B(C$_6$Cl$_5$)(C$_6$F$_5$)$_2$ (Figure 3.1). These species in THF were shown to reduce electron-deficient N-bound tosyl-based imines. In an alternative approach to sterically demanding boranes, Stephan and Erker[32] employed the 1,1-carboboration of alkynes to access alkenyl-boranes of the form R^{1}R^{2}C=C(C$_6$H$_5$)B(C$_6$F$_5$)$_2$, (Figure 3.1). These species proved effective as a catalyst for the hydrogenation of less sterically encumbered imines. The reactivity of these species varied with the steric demands, as some imines were so congested as to preclude hydride delivery.

The Soós group also exploited the bulky Lewis acids, MesB(C$_6$F$_5$)$_2$ or MesB(C$_6$F$_4$H)$_2$, in the catalytic reductions of the rings containing N atoms in a variety of quinolines.[33] Moreover Soós *et al.* demonstrated the synthetic utility of such congested catalysts, exploiting MesB(C$_6$F$_4$H)$_2$ in the reduction step of a three-step synthesis of *rac*-cuspareine (Scheme 3.14).

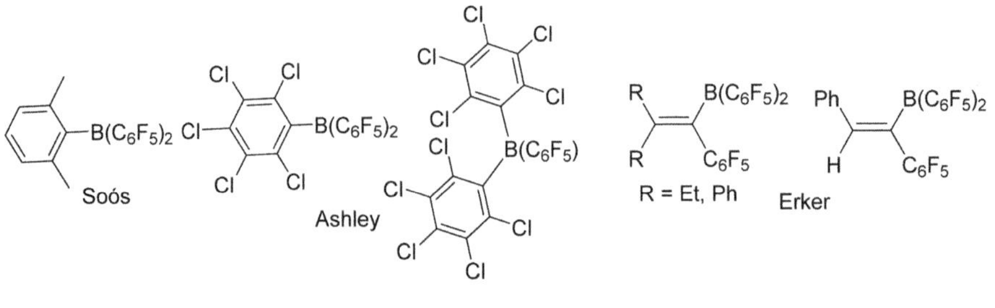

Figure 3.1   Air-stable, sterically encumbered electrophilic Lewis acids.

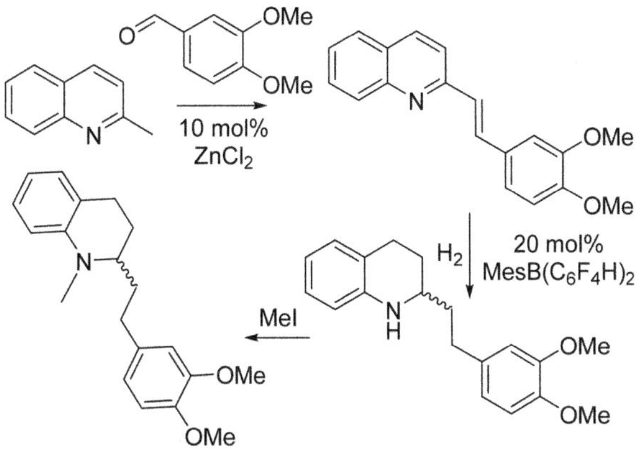

Scheme 3.14 Synthesis of *rac*-cuspareine employing an FLP hydrogenation catalyst.

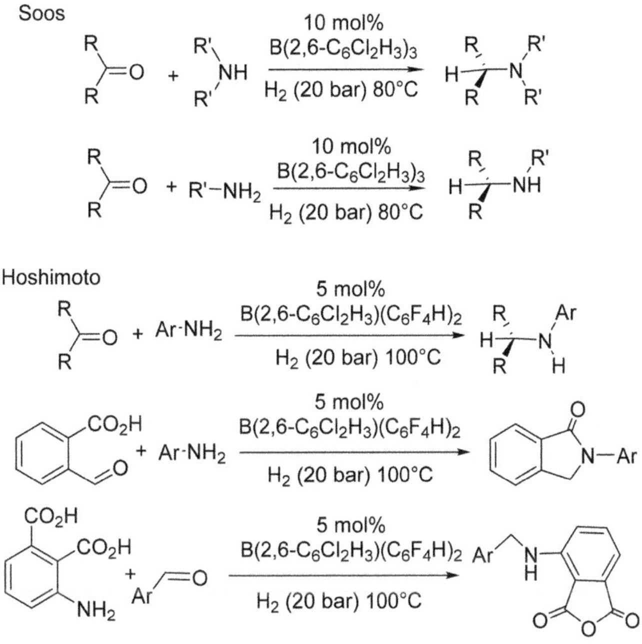

Scheme 3.15 Borane mediated tandem condensation/hydrogenation reactions.

In a further expansion of their portfolio of air- and water-stable boranes, the Soós group[34,35] prepared $B(2,6-C_6ClFH_3)_3$, $B(2,6-C_6Cl_2H_3)(2,6-C_6ClFH_3)_2$, and $B(2,6-C_6Cl_2H_3)_3$. These authors showed that the borane $B(2,6-C_6Cl_2H_3)_3$ could catalyze the condensation of aldehydes and amines, as well as the subsequent addition of dihydrogen, thus effecting the *in situ*-reduction of the generated imines to the corresponding amines (Scheme 3.15). In related efforts, Hoshimoto and coworkers[36] subsequently expanded this latter approach, using 5 mol% of the borane $B(2,6-C_6Cl_2H_3)(C_6F_4H)_2$ under optimized conditions of 100 °C and 20 bar dihydrogen pressure to mediate the condensation and reduction of aldehyde/amine combinations (Scheme 3.15). These reactions were shown to tolerate the inclusion of a range of functional groups including CO_2H, OH, $CONH_2$ and SO_2NH_2. In a similar fashion, use of carboxylic acid-derived aldehydes were condensed and reduced with amines affording a facile route to alkylated isoindolinones.

3.6 More FLP Imine Hydrogenations

Efforts to use FLP hydrogenation catalysis for the catalytic reduction of commercially viable imine substrates were also undertaken. In one effort, $B(C_6F_5)_3$ was utilized to effect the reduction of the diimines $[(C_6H_3R_2)_2N=C(Me)]_2$ (R = Me, *i*-Pr), as well as the pyridyldiimines $C_5H_3N(C(Me)=NR)_2$ (R = C_6H_4*i*-Pr, Mes, C_6H_3*i*-Pr$_2$), to the corresponding diamine derivatives (Scheme 3.16).[29] These diamines serve as precursors to N-heterocyclic carbenes and as chelating ligands for transition metals. In a more recent study, Du and coworkers applied $B(C_6F_5)_3$ to affect the reduction of a series of twenty-one symmetric and unsymmetric 1,2-diaryl-1,2-diimines. The resulting diamines were obtained in 93–99% yields of the *cis*-isomers exclusively.[37]

On the other hand, efforts to reduce pharmaceutical precursors met with mixed success. The potent analgesic and narcotic, fentanyl, was reduced in low yield, presumably a result of the presence of the donor amine center which presumably binds to boron, suppressing catalysis. However, using this metal-free protocol, an imine precursor to a potential herbicide, an analog of the antidepressant sertraline, and a precursor to an anti-cancer candidate were readily achieved (Scheme 3.17). In addition, this same study also demonstrated that under optimized conditions, imine reduction could be effected with as little as 0.1 mol% catalyst, although elevated temperatures and dihydrogen pressure were required (130 °C, 120 bar).[29]

Erker and coworkers[38] extended FLP hydrogenation to several organometallic species. For example, the conjugated metallocene species $(C_5H_4C(NR_2)CHC(CH_2)C_5H_4)M$ (M = Fe, $ZrCl_2$) were hydrogenated in the presence of $Mes_2PCH_2CH_2B(C_6F_5)_2$ and dihydrogen. The major products in both cases were of the form $(C_5H_4CH(NR_2)CHC(Me)C_5H_4)$ M (M = Fe, $ZrCl_2$) (Scheme 3.16). Interestingly, the combination of $(C_5H_4C(NC_5H_8Me_2)$ $CH_2C(Me)C_5H_4)Fe$ with $B(C_6F_5)_3$ has more recently been shown to mediate the

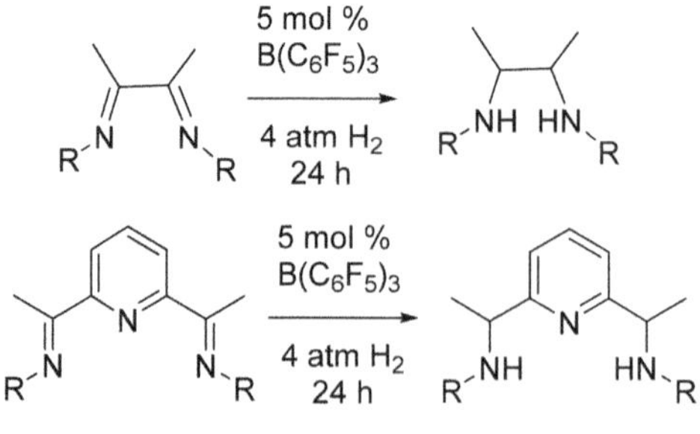

Scheme 3.16 FLP hydrogenations of diimine substrates.

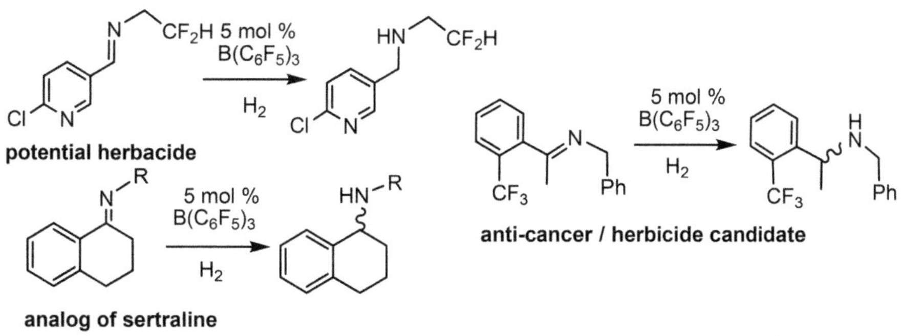

Scheme 3.17 FLP reductions of pharmaceutical precursors.

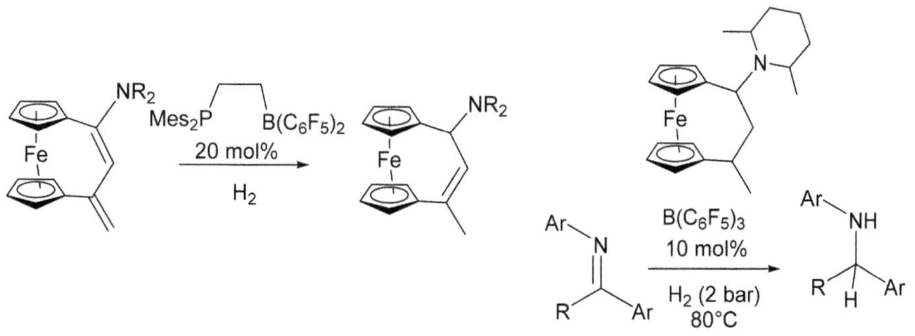

Scheme 3.18 FLP reduction of organometallic substrate and application in FLP hydrogenations.

hydrogenation of a series of aryl ketimines using 2 bar dihydrogen at 80 °C (Scheme 3.18).[39]

In 2016, Krempner and coworkers[40] examined the use of weaker Lewis acids in the activation of dihydrogen and catalysis. These authors demonstrated that BPh_3 and $HBMes_2$ could affect the heterolytic cleavage of hydrogen in the presence of stronger bases such as $(C_4H_8N)_3PNt\text{-}Bu$ and $N(CH_2CH_2Ni\text{-}Pr)_3P$. This latter base also activated dihydrogen in the presence of (C_6H_{13})-BBN and BEt_3. These combinations were evaluated as catalysts for the hydrogenation of PhCH=NPh at 25–60 °C using 50–100 bar of dihydrogen pressure with a 5 mol% catalyst loading. Of these, $[(C_4H_8N)_3PNt\text{-}Bu/HBPh_3$ and $N(CH_2CH_2Ni\text{-}Pr)_3P/(C_6H_{13})$-BBN proved most efficient.

In a more recent expansion of the scope of FLP reductions, Oestreich and co-workers described the $B(C_6F_5)_3$-mediated hydrogenation of *O*-alkyl and *O*-silyl oxime ethers to *N*-monosubstituted hydroxylamines (Scheme 3.19). The N–O bond remains through the catalysis but is readily hydrolyzed during workup to give the corresponding primary amines.[41] Similarly, aldoximine ethers, as well as ketone- and aldehyde-derived hydrazones were reduced[42] and deprotection converted phthaloyl-protected hydrazones to acetyl-substituted hydrazines in excellent yields.[41,42]

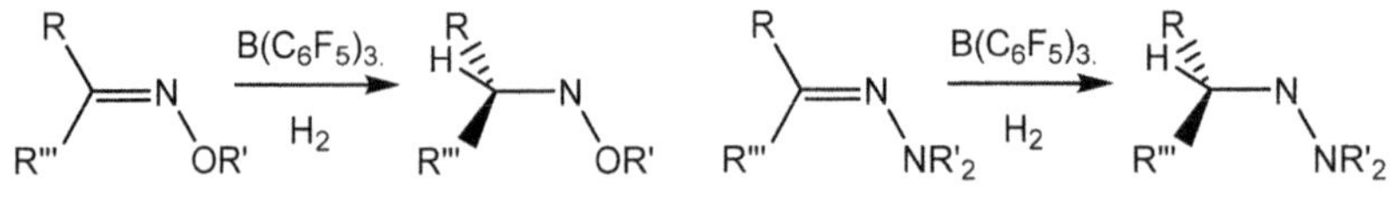

Scheme 3.19 Oxime and hydrazine reductions.

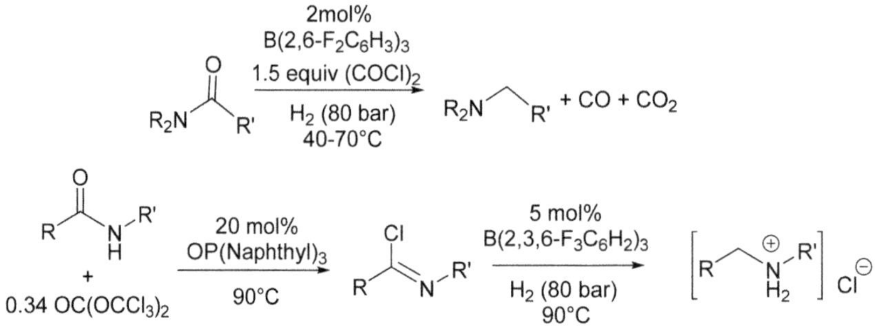

Scheme 3.20 FLP reduction is that of the CO fragment in an amide.

A particularly challenging reduction is that of the amide. Paradies and coworkers[43] developed an innovative approach to solving this problem. Treatment of amides with $(COCl)_2$ in the presence of the borane, $B(2,6\text{-}F_2C_6H_3)_3$ prompts the *in situ* conversion of the amide to a chloride iminium cation with the release of CO and CO_2. Subsequent reaction with hydrogen results in the conversion to the corresponding ammonium chloride. While such conversions only employed 2 mol% of the borane, it did require the use of 80 bar of dihydrogen at 40–70 °C. Nonetheless, these reductions were applied to a series of amide precursors and yields were typically in the range of 63–99% (Scheme 3.20).

In a very recent publication, Paradies and coworkers[44] extended the above protocol to secondary amides. Thus, treatment of $RC(O)NHR'$ with $OC(OCCl_3)_2$ in the presence of a catalytic amount of $(Napthyl)_3PO$ mediated the conversion of the amide to the chloro-imine $RC(Cl)=NR'$ (Scheme 3.20). Subsequent hydrogenation mediated by 5 mol% of $B(2,3,6\text{-}F_3C_6H_2)_3$ generated the salt $[RCH_2NH_2R']Cl$ (Scheme 3.20). In general, the optimized conditions involve reactions at 90 °C, with the initial 15 h of the catalysis generating the chloro-imine and a further 20 h using 80 bar of dihydrogen pressure. Yields of the products ranged from 61–86%. Like conventional imine reductions, the mechanism of hydrogenation of the chloro-imines was proposed to operate by both simple and autocatalytic cycles. A computational study of this system by Grimme *et al.*[45] suggests the role of borane and chloride in the activation of dihydrogen.

3.7 Borane-mediated Hydrogenations of N-based Heterocycles

The scope of reductions was further broadened to include N-based heterocycles. In our initial report, the nitrogen containing heterocycles including substituted quinolines, phenanthroline and acridine derivatives, were hydrogenated under mild conditions

(25 °C) using 5–10 mol% of $B(C_6F_5)_3$ as the catalyst under 4 bar of dihydrogen.[46] In this fashion, acridine was reduced to dihydroacridine, substituted quinolines were reduced to tetrahydro-quinolines, and phenanthroline was reduced to tetrahydro-phenanthroline (Scheme 3.21). It is noteworthy that the latter reduction proceeded even though the $B(C_6F_5)_3$–phenanthroline adduct could be isolated in stoichiometric reaction in the absence of dihydrogen (Scheme 3.21).

Building on the above results, this protocol was extended to a series of pyridines affording the corresponding piperidinium salts. Similarly, quinolines, acridine and quinoxaline derivatives were fully reduced to afford the corresponding hydridoborate salts (Scheme 3.22).[47] In the case of benzoquinoline, only the N-containing ring and the adjacent ring were reduced, while the disparate arene ring remained unaltered (Scheme 3.22).[47] In these cases, the formation of ammonium salts sequestered the Lewis acid as the counterion, halting catalysis.

While these latter N-heterocycle reductions are not strictly catalytic in nature, they do lead to the repetitive consumption of dihydrogen. Indeed, in these cases, catalytic turnover is halted by the activation of dihydrogen by the generated amine and borane, thus sequestering the borane from the reaction mixture. A subsequent computational study

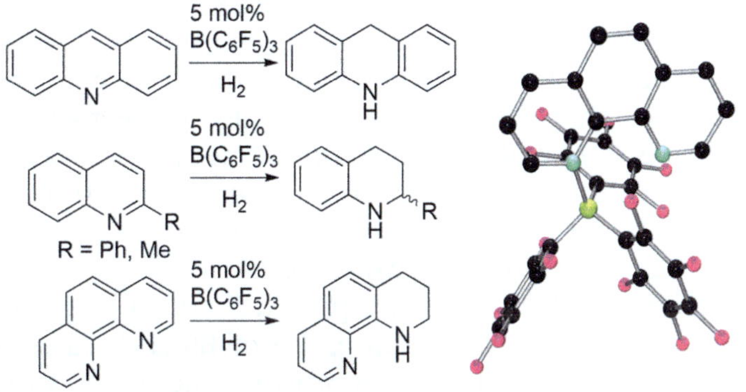

Scheme 3.21   FLP hydrogenations of nitrogen-based heterocycles; structure of $B(C_6F_5)_3$–phenanthroline adduct.

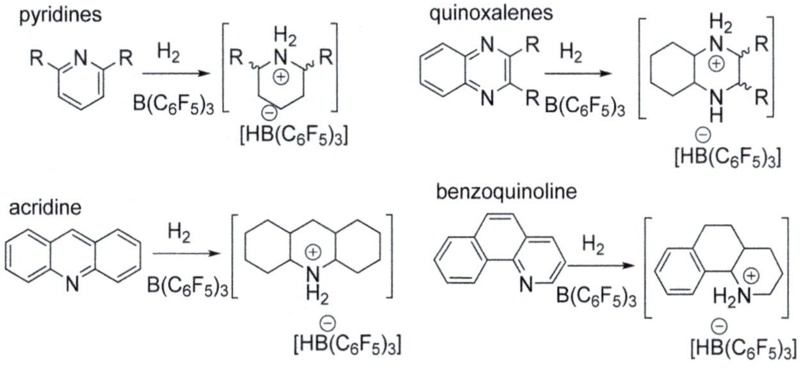

Scheme 3.22   Stoichiometric reductions of N-heterocycles.

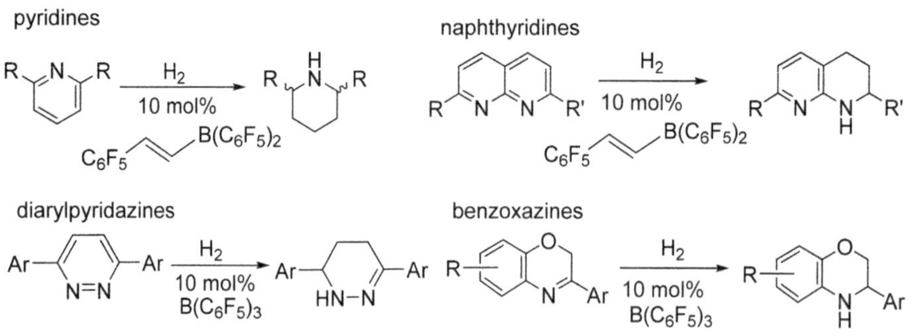

Scheme 3.23 Examples of FLP reductions of N-heterocycles.

by Li and co-workers showed the conventional FLP mechanism involving dihydrogen activation by the borane and pyridine is the rate-determining step.[48]

The above stoichiometric reductions of substituted pyridines gave a mixture of diastereomers, consistent with the ability of $B(C_6F_5)_3$ to epimerize chiral centers *alpha* to nitrogen.[49] In a subsequent related effort, Du and co-workers[50] demonstrated that related tetrahydronaphthylamines could be obtained using 10 mol% of $B(C_6F_5)_3$ and dihydrogen at 60 °C in the reduction of naphthylamines.

In a 2013 study, Du *et al.*[51] demonstrated catalytic reduction of substituted pyridines using *in situ* generated boranes derived from the hydroboration of olefins with Piers' borane, $HB(C_6F_5)_2$. This catalysis provided a highly effective route to a series of *cis*-piperidines (Scheme 3.23). Subsequently, this same catalyst was also employed to reduce a range of 3,6-disubstituted naphthyridines to 1,2,3,4-tetrahydro-1,8-naphthyridines (Scheme 3.23).[52] Under catalytic conditions using $B(C_6F_5)_3$ 3,6-diarylpyridazines were reduced to 1,4,5,6-tetrahydropyridazine derivatives[53] (Scheme 3.23), although sterically unencumbered dialkylpyridazines proved inert under similar reaction conditions. Du's group[54] also showed that 2.5 mol% of $B(C_6F_5)_3$ catalyzed the hydrogenation of 3-substituted 2H-1,4-benzoxazines providing access to a variety of range of 3,4-dihydro-2H-1,4-benzoxazines (Scheme 3.23).

3.8 Intramolecular Catalyst Variations

As early results recognized that intramolecular FLP catalysts are pre-organized for interaction with dihydrogen, thus speeding catalysis, further studies probed structural variation of intramolecular FLPs. One approach was to vary the linker between the boron and phosphorus donor in B/P FLPs (Figure 3.2). Variants included the ferrocene-linked FLP, $Fe(C_5H_4PMes_2)(C_5H_4CH_2CH_2B(C_6F_5)_2)$,[55] as well as the carbon-linked FLPs, $C_5H_8(PMes_2)B(C_6F_5)_2$,[56] $Mes_2PCH_2CH(Me)CH_2B(C_6F_5)_2$,[57] $Mes_2P(H)CH_2CH_2C(=C(SiMe_3)C(=CH_2)Me)B(C_6F_5)_2H$[58] and $Mes_2PHCH_2C(=CHSiMe_3)B(C_6F_5)_2H$.[59] While all these systems could be employed in enamine hydrogenations, the activities were unexceptional.

Other efforts targeted the B/N intramolecular FLP catalysts. For example, Mitzel *et al.*[60,61] developed a series of new B/N species, such as $PhN(CH_2CH_2CH_2B(C_6F_5)_2)_2$, that catalytically reduced the silyl-enol ether $Ph(Me_3SiO)C=CH_2$ in 42% yield under 6 bar of

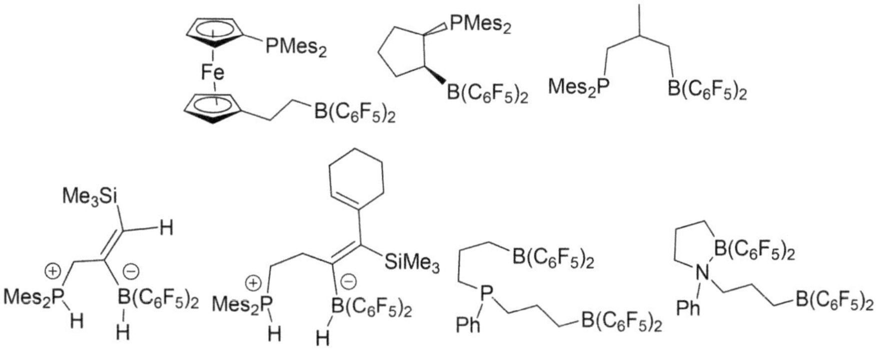

Figure 3.2 Other Intramolecular FLP catalysts.

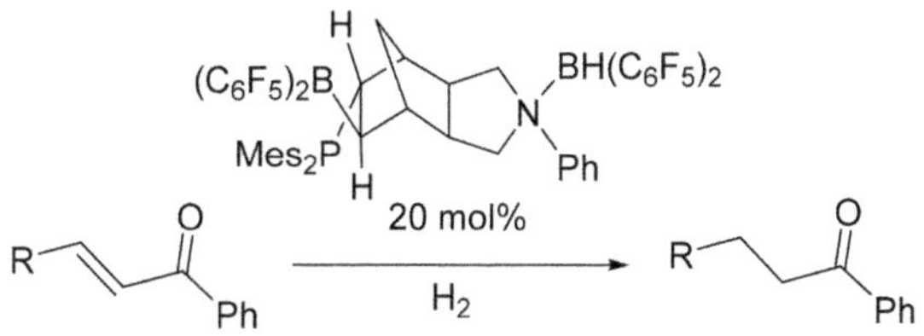

Scheme 3.24 A unique intra-intermolecular FLP in enone hydrogenation.

dihydrogen (Figure 3.2). Interestingly, this species was unreactive in efforts to reduce $Ph_2C=CH_2$ or enamines.

More recently, the Erker group has reported[62] an interesting intramolecular FLP that also contains an intermolecular acid–base adduct site. This species, derived from a camphor derivative, $(C_6F_5)_2BHN(Ph)(CH_2)_2C_7H_8(PMes_2)(B(C_6F_5)_2)$, reacted with dihydrogen to give the intramolecular zwitterion $PhN(CH_2)_2C_7H_8(PHMes_2)(BH(C_6F_5)_2)$. While this latter species was inactive in the catalytic hydrogenation of imines, 20 mol% of this zwitterion mediated the reduction of a series of enones of the form $RCH=CHC(O)Ph$ at room temperature under 2 bar of dihydrogen (Scheme 3.24).

Multiple groups attempted to prepare the geminal FLP t-$Bu_2PCH_2B(C_6F_5)_2$; however, these were unsuccessful affording instead the cyclized zwitterion t-$Bu_2PCH_2BF(C_6F_5)$ (C_6F_4) (Figure 3.3).[63–65] In contrast, the geminal FLP t-$Bu_2PCH_2BPh_2$ was prepared by Lamertsma and coworkers[66] and was shown to react with dihydrogen affording the zwitterion $[t$-$Bu_2PHCH_2BHPh_2]$ (Figure 3.3). Although the Lewis acidity at boron is less than that in the classic Erker intramolecular FLPs, the phosphine is more basic, consistent with the requirement for a high cumulative acidity and basicity for dihydrogen activation. The related geminal B/P FLP, t-$Bu_2PCH_2B(C_6H_3(CF_3)_2)_2$, was prepared by Wagner and co-workers (Figure 3.3)[67] and was used to mediate the reduction of t-$BuN=CHPh$. More recently, Erker *et al.* have prepared the geminal FLP variant, $((C_6F_5)_2B)((R_3C_6H_2)HP)C=CH(t$-$Bu)$ (R = t-Bu, Me) (Figure 3.3), which contains a primary phosphine fragment.[68] However, this species has not been evaluated in catalysis. Similarly, the Erker

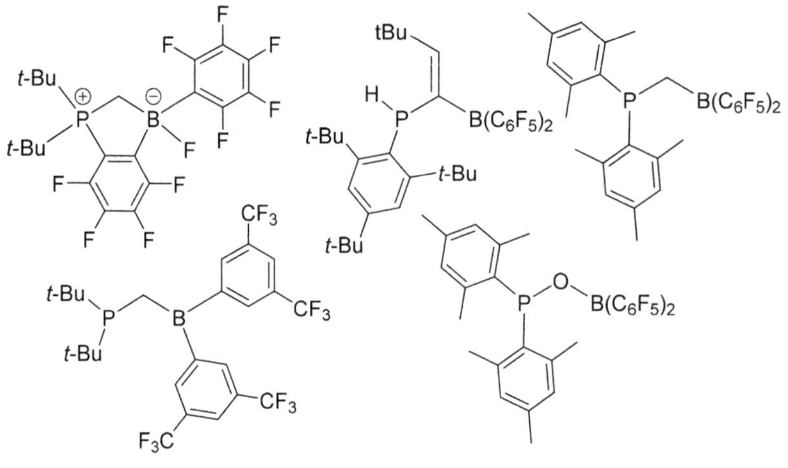

Figure 3.3 Geminal intramolecular P/B FLPs.

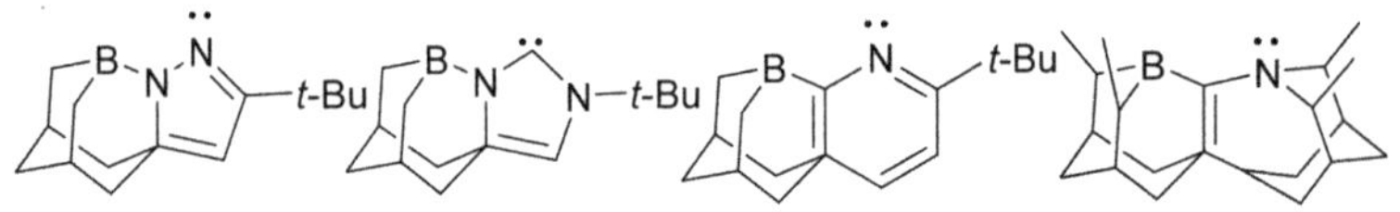

Figure 3.4 Hypothetical geminal N/B FLPs.

group[69] developed a synthesis pathway to the geminal FLP, $Mes_2PHCH(Me)BH(C_6F_5)_2$. 10 mol% of this salt was shown to successfully undergo hydrogenation of an imine, enamine, silyl-enol ether, and the N-ring of a quinoline derivative, at near quantitative conversions after 36 h. In 2018, the geminal species $Mes_2POB(C_6F_5)_2$ in which the linking atom is oxygen, was reported.[70] This species was also shown to react with dihydrogen, although 50 bar of dihydrogen and 50 °C were required to obtain the zwitterion $[Mes_2P(H)O(H)B(C_6F_5)_2]$.

A subsequent computational study[71] of a series of group 15/13 geminal systems revealed the kinetic barrier to dihydrogen activation increases down the group 15 (N < P < As < Sb) regardless of the nature of the group 13 element. Further computational studies have probed the hypothetical system of the form $H_2NCH_2BR_2$,[72] while Wang *et al.*[73] predicted that a series of intramolecular geminal B/N FLPs would also activate dihydrogen efficiently (Figure 3.4). These latter systems are food for thought, although protocols for the syntheses of such species are currently unknown.

3.9   Hydrogenation of Olefins

As described above, early results from the Erker group described the formal reduction of C=C double bonds in enamines and silyl enol ethers. In a 2010 paper, Soós described the FLP reductions of a conjugated olefinic bond.[33] Specifically, these authors reported the conversion of carvone $(CH_2=C(Me)C_6H_5(Me)O)$ to $(CH_2=C(Me)C_6H_7(Me)O)$ (Scheme 3.25) and the reduction of MeCH=CHCH=Nt-Bu to the saturated amine BuN(H)t-Bu.

In 2012, a Stephan/Erker collaboration employed a series of sterically encumbered alkenyl boranes and DABCO as FLPs to effect the C=C bond reduction in enones.[32] Similarly, subsequent work by Ashley and co-workers employed the borane $B(C_6Cl_5)$ $(C_6F_5)_2$ as a catalyst in THF to hydrogenate $CH_2=CHC(O)On$-Bu.[31] In addition, the Erker group described related enone reductions employing the FLP derived from $CpFe(C_5H_4$ $PMes_2)/B(C_6F_5)_3$[55] and the Paradies group showed that enones could be consecutively hydrosilylated and hydrogenated using bis-phosphines and $B(C_6F_5)_3$ in the presence of silane and dihydrogen (Scheme 3.26). A similar approach was subsequently applied to the hydrogenation of allyl-silanes derived from the hydrosilylation of fulvenes.[74]

However, it was a 2012 collaborative report from the groups of Paradies and Stephan[75] that demonstrated the FLP reductions of un-activated terminal olefins. This team examined the use of an FLP derived from $(C_6F_5)Ph_2P$ and $B(C_6F_5)_3$ in the activation of dihydrogen, and subsequently used this FLP to affect the reduction of a series of 1,1-disubstituted olefins under relatively mild conditions of 25–50 °C and 4 bar of dihydrogen to afford the corresponding alkanes in high yields. Nonetheless, these reactions were not fast, typically taking from 12–240 h to go to completion depending on the steric and electronic features of the substituents on the olefin.

The remarkable ability of this FLP to heterolytically cleave dihydrogen at temperatures as low as −60 °C to −80 °C generated a highly acidic phosphonium cation that was able to protonate olefins. The resulting carbocation was sufficiently Lewis acidic to

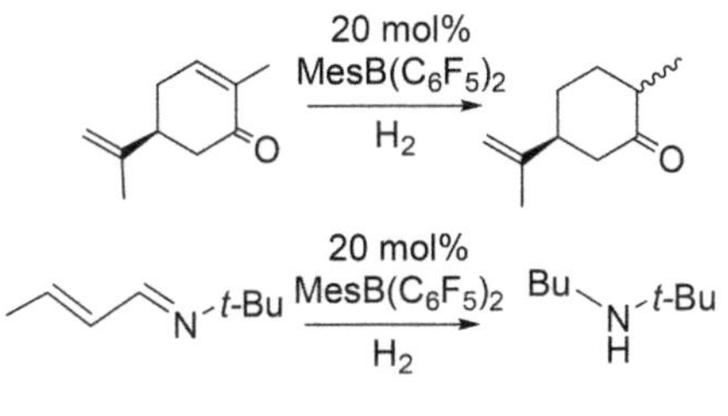

Scheme 3.25 FLP reduction of carvone and MeCH=CHCH=Nt-Bu.

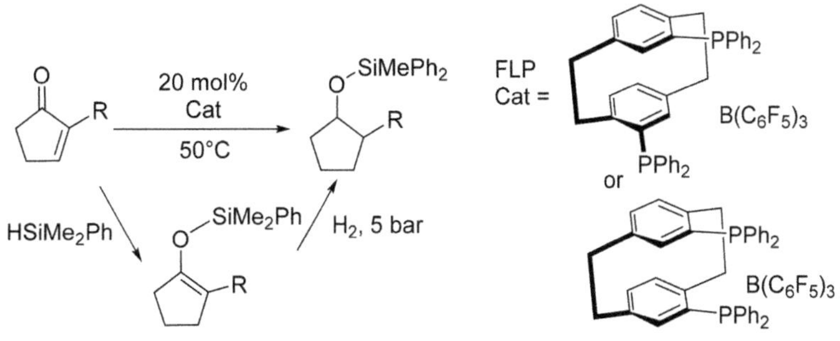

Scheme 3.26 Consecutive hydrosilylation/hydrogenation of enones by FLPs.

capture hydride from the borohydride anion resulting in the net reduction of the olefin (Scheme 3.27). The FLP derived from P(1-naphthyl)$_3$ and B(C$_6$F$_5$)$_3$ was also an effective hydrogenation catalyst for such olefins. In the presence of the base, Ph$_2$NMe, the transient carbocation was intercepted to give Ph$_2$MeC(C$_6$H$_4$)NMePh *via* electrophilic substitution at the carbon atom *para*- to the carbon cation center, thus confirming the sequence of proton and hydride delivery in these olefin reductions.

Noting the known acidity of Jutzi's acid,[76] [(Et$_2$O)$_2$H][B(C$_6$F$_5$)$_4$], we probed the reactivity of the simple adduct (Et$_2$O)•B(C$_6$F$_5$)$_3$.[77] In toluene, this adduct established an equilibrium between free Lewis acid, ether, and the adduct. From this mixture, heterolytic cleavage of dihydrogen would generate the salt, [[(Et$_2$O)$_2$H][HB(C$_6$F$_5$)$_3$], which is both a strong acid and a hydride source. Although this species could not be spectroscopically observed, its formation was confirmed using HD gas, as isotopic scrambling was observed, affording a mixture of HD, dihydrogen and dideuterium. This observation confirmed the transient generation of the salt, [(Et$_2$O)$_2$H(D)][H(D)B(C$_6$F$_5$)$_3$], which on the random reaction of the cation with the anion gave the observed mixture of isotopomers of dihydrogen. However, as in the phosphine case above, in the presence of olefin, the protonated ether is sufficiently acidic to protonate a 1,1-disubstituted olefin, providing an alternative and remarkably simple FLP catalysts for the hydrogenation of olefins (Scheme 3.28), *i.e.*, (Et$_2$O)•B(C$_6$F$_5$)$_3$.

In a subsequent study, Autrey and coworkers[78] examined the reaction enthalpy and rate of diphenylethylene hydrogenation by the FLP B(C$_6$F$_5$)$_3$/Et$_2$O. They revealed the rate of conversion to alkane showed a linear dependence on both free Et$_2$O and borane

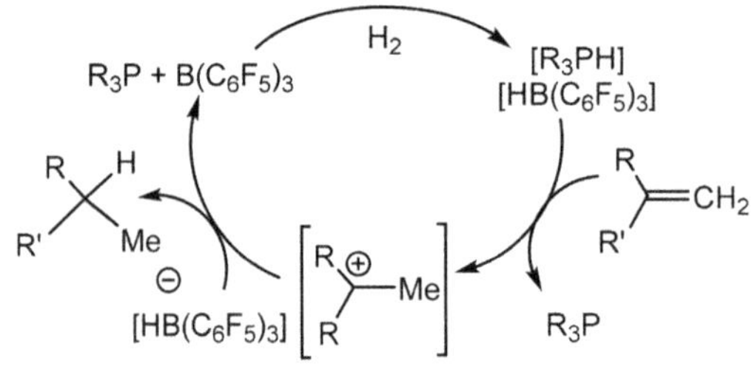

Scheme 3.27 FLP hydrogenation of 1,1-disubstituted olefins.

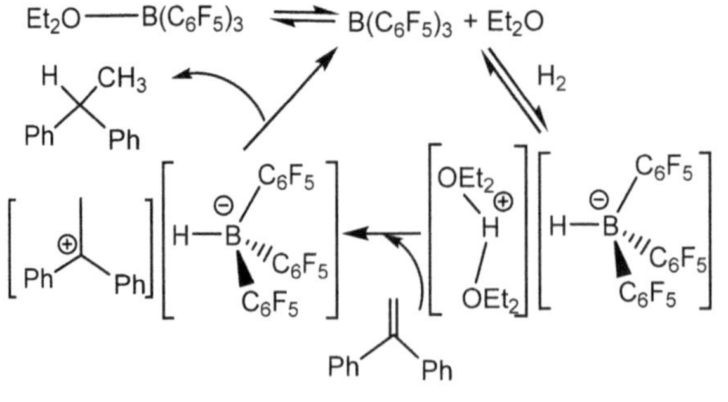

Scheme 3.28 Mechanism of hydrogenation of 1,1-diphenylethene mediated by Et$_2$O/B(C$_6$F$_5$)$_3$.

concentrations. The reaction enthalpy was determined to be $\Delta H = -116 \pm 4$ kJ mol^{-1}, while the equilibrium of the ether–borane adduct had thermodynamic parameters of $\Delta H = -54.6 \pm 3.3$ kJ mol^{-1} and $\Delta S = -154 \pm 13$ J mol^{-1} K^{-1}.

Also in 2012, Alcarazo and co-workers accomplished the catalytic hydrogenation of the electron-poor allenes $R_2C=C=C(CO_2Et)_2$, and alkenes, $RCH=C(CO_2Et)_2$ using 5 mol% DABCO and $B(C_6F_5)_3$ as the catalysts under dihydrogen (1 bar).[79] In the case of these electron-deficient alkenes, in contrast with the electron-rich olefins above, these authors proposed a reaction pathway that involved initial hydride delivery to the olefinic carbon followed by protonation at oxygen. Then, a subsequent proton migration afforded the reduced alkane. This pathway was supported by labeling experiments (Scheme 3.29). The next year, Alcarazo's group studied similar reductions using a series of fluorinated boranes as the Lewis acid component of the FLP.[80] They showed that while dihydrogen cleavage was accelerated by enhanced Lewis acidity at boron, the subsequent delivery of hydride was slowed. Ultimately, $B(2,4,6\text{-}C_6H_2F_3)_3$ was deemed optimal among the boranes used for these reductions.

Expanding the scope of electron-deficient olefins, Paradies' group[81] examined the hydrogenation of a series of nitro-olefins and acrylates using $(THF)\cdot B(2,6\text{-}C_6F_2H_3)_3$ and 2,6-lutidine or collidine as the FLP catalyst at room temperature and under 4 bar dihydrogen. These efforts illustrated that judicious selection of the FLP afforded tolerance of the functional groups in these olefinic substrates. Interestingly, the reaction of the olefin $PhCH=CH(NO_2)$ with $t\text{-}Bu_3P$ and $B(2,6\text{-}C_6F_2H_3)_3$ resulted in the addition of the FLP to an olefin to give the zwitterion $(t\text{-}Bu_3P)PhCHCH=NO(OB(2,6\text{-}C_6F_2H_3)_3)$ (Scheme 3.30). The corresponding reaction with tetramethylpiperidine and dihydrogen (4 bar) affords the hydrogen-bonded ion pair, $[C_6H_6Me_4NH][PhCH_2CHNO(OB(2,6\text{-}C_6F_2H_3)_3)]$ (Scheme 3.30), whereas tetramethylpiperidine and dihydrogen afforded the hydrogen-bonded ion pair, $[C_6H_6Me_4NH][PhCH_2CHNO(OB(2,6\text{-}C_6F_2H_3)_3)]$. This latter result supports a mechanism of hydrogenation in which hydride is delivered initially to the olefin. Again, this stands in contrast to the order of delivery of proton and hydride for other FLP reductions. These results were further supported by the work of Berrioni and co-workers[82] who studied the kinetics of the reactions of a family of salts derived from

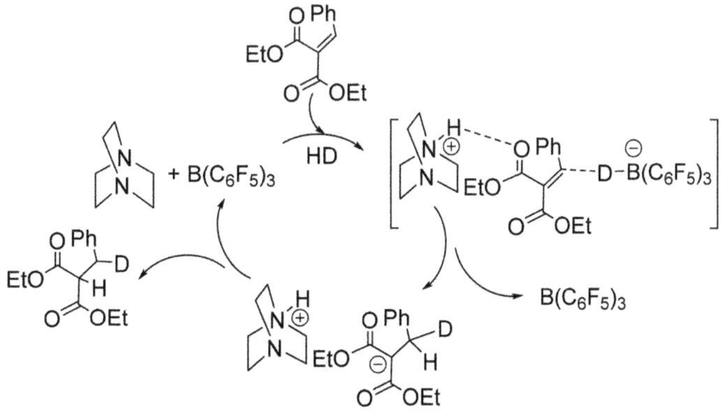

Scheme 3.29 Alcarazo's proposed reaction pathway for the reduction of $RCH=C(CO_2Et)_2$.

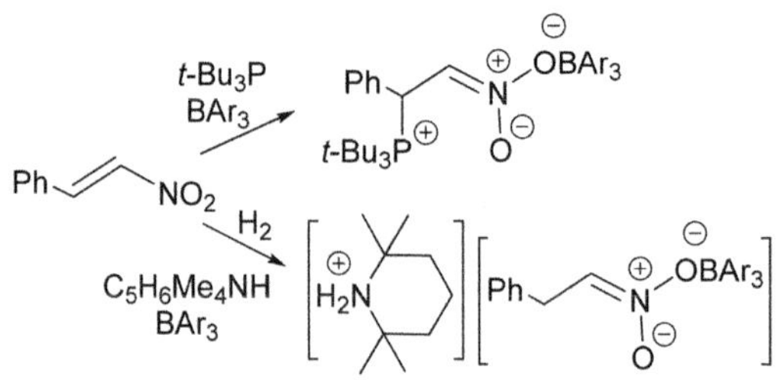

Scheme 3.30 Reactions of PhCH=CH(NO$_2$) with FLPs.

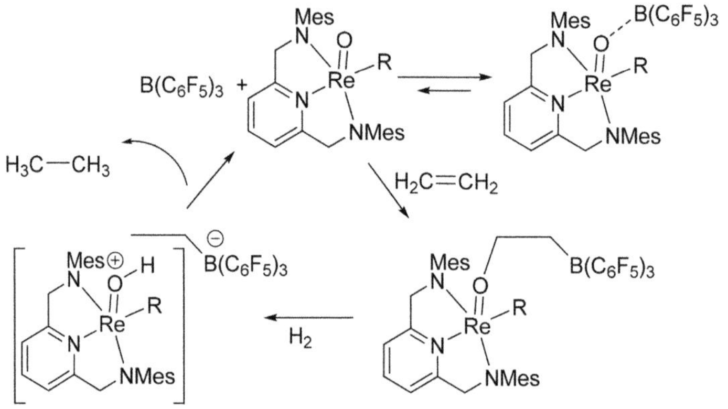

Scheme 3.31 FLP hydrogenation of ethylene by the ReO/B FLP.

the reaction of an FLP with dihydrogen and their subsequent reaction with electron-deficient olefins. The rate of hydride transfer to the electron-poor olefins was shown to be independent of the counterion but was impacted by the hydricity of the triaryl borane, consistent with the observations described by Paradies.

Broadening the catalysts for olefin reduction, the Ison group[83] reported the use of the rhenium complex C$_5$H$_3$N(CH$_2$NMes)$_2$Re(O)Ph as the base partner of B(C$_6$F$_5$)$_3$ in an FLP. While the borane was shown to form an adduct with the oxygen of the ReO fragment, the presence of an equilibrium mixture provided access to the FLP, which was capable of reaction with dihydrogen. In that sense, this system behaves like the aforementioned (Et$_2$O)•B(C$_6$F$_5$)$_3$ system. Ison went on to show that 5 mol% of the metal complex with 10 mol% of borane mediated the hydrogenation of a series of olefins and dienes under 3.4 bar of dihydrogen at 100 °C. After 16 h, the yields of the reduced products ranged from 82–99%, although cyclohexene was only reduced in 12% yield. A detailed examination of the mechanism (Scheme 3.31) demonstrated that it was unambiguously distinct from the traditional transition metal-catalyzed mechanisms, which require coordination of the substrates to the metal center. A subsequent study[84] optimized this catalytic system, demonstrating that increasing the steric demands of the substituents on the N-atoms favors FLP formation. In addition, the reaction rates were improved with the use of B(C$_6$F$_5$)$_3$, or Al(C$_6$F$_5$)$_3$ allowing this optimized system to effect hydrogenation of un-activated olefins at room temperature. In contrast, a recent study[85] of the Re-nitride complex $S(CH_2CH_2S)_2Re(N)PPh_3$ and $B(C_6F_5)_3$ showed that it was ineffective for catalytic hydrogenation of olefins, even at 100 °C and 3.5 bar of dihydrogen.

In exploring the fundamental details of the interaction of boranes with dihydrogen, Nikonov and coworkers[86] showed that $B(C_6F_5)_3$ could not activate dihydrogen on its own, but that $HB(C_6F_5)_2$ affected H/D exchange in reactions with D_2. This finding prompted an alternative strategy to metal-free hydrogenation of olefins. To this end, Wang, Li, and coworkers[87] used 20 mol% of $HB(C_6F_5)_2$ as a catalyst at 140 °C under 6 bar of dihydrogen pressure for the reduction of the terminal, di- and tri-substituted olefins. The proposed catalytic cycle begins with the hydroboration of the substrate followed by hydrogenolysis, which regenerates the catalyst affording the reduced product. The authors describe the hydrogenolysis as a σ-bond metathesis reaction, although one can view this as the activation of dihydrogen across the polar B–C bond, in which the boron is Lewis acidic, and the carbon is basic (Scheme 3.32), akin to that described for $R_2PB(C_6F_5)_2$ (R = Cy, *t*-Bu) (see Section 2.11).

In a more recent paper, Melen and co-workers[88] have incorporated FLP mediated olefin reductions into a tandem process. These authors used a Lewis acid to promote the *in situ* generation of enamines derived *via* aza-Morita–Baylis–Hillman adducts, and then undergo subsequent hydrogenation of the activated olefinic bond (Scheme 3.33). In this fashion, a series of β-amino acid derivatives were prepared in diastereomeric ratios ranging from 55:45 to 90:10.

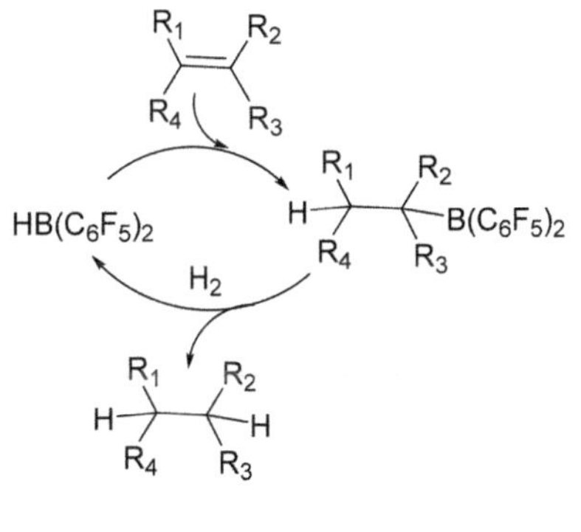

Scheme 3.32   Catalytic reduction of olefins using Piers' borane.

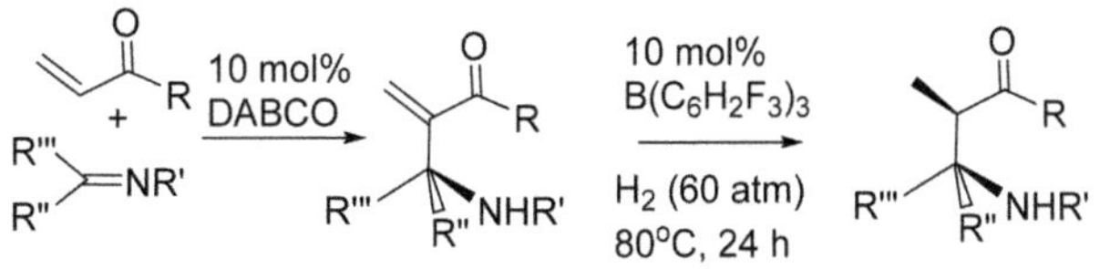

Scheme 3.33   Borane mediated tandem condensation/hydrogenation reactions.

3.10   FLP Hydrogenation of Alkynes

In 2011, efforts to reduce the activated alkyne derivative PhCCC(O)t-Bu employing the alkenylborane t-BuCH=C(C$_6$H$_5$)B(C$_6$F$_5$)$_2$[89] afforded the *trans* enone PhCH=CHC(O)t-Bu as the major product, with a small amount of the corresponding ketone, PhCH$_2$CH$_2$C(O)t-Bu, as a minor product. However, it was the 2013 report from Repo and co-workers,[90] that demonstrated a general approach to the FLP reduction of alkynes. This innovation exploited intramolecular B/N FLPs, to hydrogenate a wide variety of internal alkynes to the corresponding *cis* olefins. The isolation of intermediates, isotopic label-ing experiments, in addition to computational studies, revealed that the initial step involves the reaction of (C$_6$H$_4$)(NMe$_2$)(B(C$_6$F$_5$)$_2$) with dihydrogen to liberate C$_6$F$_5$H and generate C$_6$H$_4$NMe$_2$(B(C$_6$F$_5$)H). This latter product reacts with the alkyne *via* a hydrobo-ration reaction while subsequent heterolytic cleavage of dihydrogen prompts intramo-lecular proto-deborylation, liberating the *cis*-alkene product and regenerating the FLP, C$_6$H$_4$NMe$_2$(B(C$_6$F$_5$)H) for further reactions (Scheme 3.34). The regioselectivity observed for these reactions is attributed to the hydroboration step in the mechanism.

In a subsequent report, these authors[91] examined the related, B/N FLP C$_6$H$_4$(NMe$_2$)BCl$_2$. This species also afforded catalytic reduction of several alkynes to give *cis*-olefin products, although pentamethylpiperidine was used as a promoter. Nonetheless, the observed regioselectivity suggested a mechanism like that seen for the C$_6$F$_5$ analog, although the precise details of the mechanism were not confirmed.

In 2015, the Du group[92] also examined the hydrogenation of alkynes. In their case, they used HB(C$_6$F$_5$)$_2$ in the presence of C$_6$F$_5$CH=CH$_2$ as the catalyst. This initial com-bination prompts *in situ* hydroboration of the alkyne to give RCH=CR'B(C$_6$F$_5$)$_2$. Sub-sequent hydrogenolysis of the borylated alkene gives the *cis*-alkene product, freeing HB(C$_6$F$_5$)$_2$ for further reaction with an alkyne. Interestingly, once the alkyne is depleted, the HB(C$_6$F$_5$)$_2$ is available to affect the isomerization of the Z-alkene to the E-alkene in a hydroboration/retro-hydroboration cycle (Scheme 3.35).

A further expansion of FLP hydrogenations of terminal alkynes was achieved by Gell-rich.[93] In a cleaver approach, the authors prepared a unique intramolecular FLP from the combination of a pyridine derivative with Piers' borane C$_5$H$_3$(t-Bu)N(OB(C$_6$F$_5$)$_2$)N.

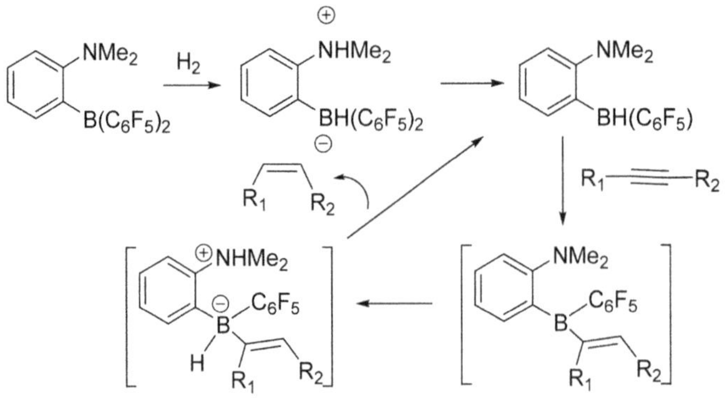

Scheme 3.34   Regioselective FLP reduction of alkynes.

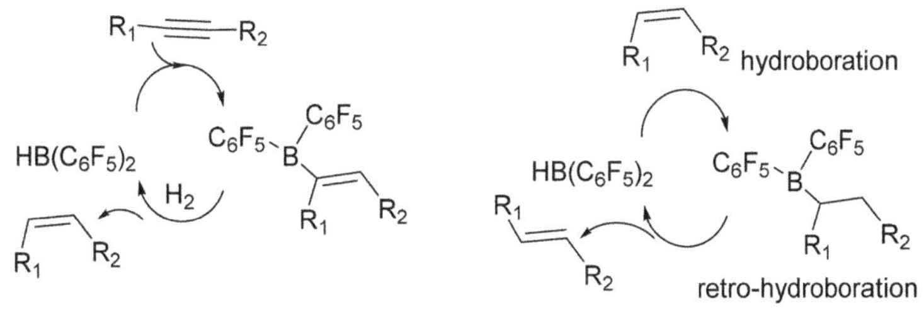

Scheme 3.35 Reduction alkyne and isomerization of *Z*-olefin to *E*-olefin.

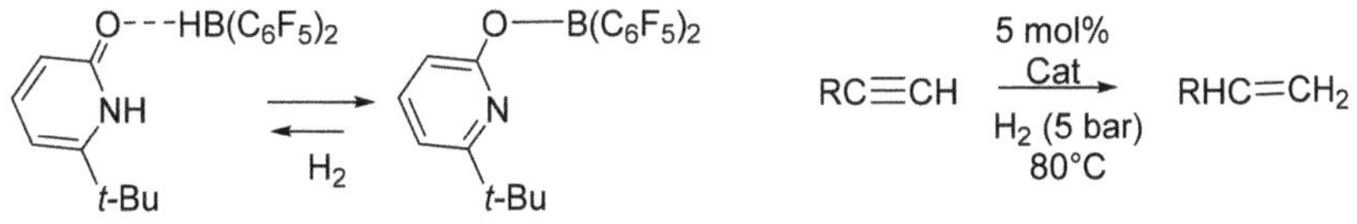

Scheme 3.36 FLP reduction of terminal alkynes.

This species reacted reversibly with dihydrogen under mild conditions as this species behaves as a pyridine adduct of Piers' borane. This feature also proved to facilitate the hydrogenation of terminal alkynes to the corresponding *cis* olefin (Scheme 3.36).[94] A subsequent computational study[95] probed variants in which the donor atom was S or N, suggesting that the sulfur-based analog exhibits the smallest activation barrier and thus is a promising synthetic target for further study.

3.11 Aromatic Reductions

Efforts to reduce polycyclic aromatic hydrocarbons using FLPs have also been reported.[96] Exploiting the FLP derived from $B(C_6F_5)_3$ and $Ph_2P(C_6F_5)$, the catalytic addition of one equivalent of hydrogen to several anthracenes and tetracene derivatives was achieved under high pressures of 102 bar at 80 °C after 10–48 h (Scheme 3.37). It is noteworthy that Ashley *et al.*[31] employed the air-stable borane $B(C_6Cl_5)(C_6F_5)_2$ to mediate the corresponding reduction of anthracene to dihydroanthracene under somewhat milder conditions.

Further expansion of FLP aromatic reductions came from a serendipitous finding. In exploring the activation of dihydrogen with readily accessible bases, reactions of a series of sterically encumbered secondary anilines of the form ArN(R)H with $B(C_6F_5)_3$ with dihydrogen were probed. At ambient temperatures, simple heterolytic cleavage of dihydrogen afforded the expected anilinium salts, $[ArN(R)H_2][HB(C_6F_5)_3]$. For example, combinations of the amine, *t*-BuNHPh, and $B(C_6F_5)_3$ at 25 °C under dihydrogen (1 bar) gave the salt $[t\text{-BuNH}_2Ph][HB(C_6F_5)_3]$. However, additional heating of the reaction mixture to 110 °C for 24–96 h resulted in the reduction of the phenyl ring to form the salt, $[t\text{-BuNH}_2Cy][HB(C_6F_5)_3]$ (Scheme 3.38).[97] This formation of the cyclohexylammonium hydridoborate salts provides a surprising and unique metal-free aromatic reduction

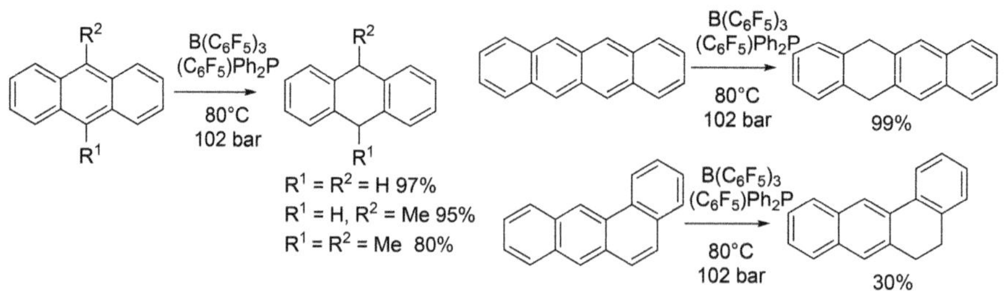

Scheme 3.37 FLP reduction of polyaromatic species.

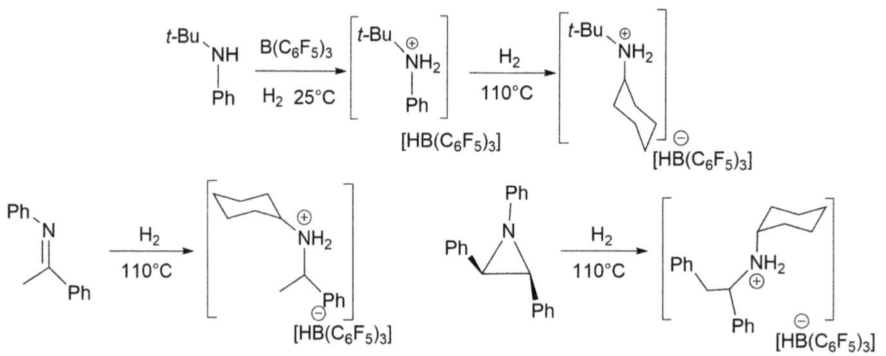

Scheme 3.38 Examples of FLP reduction of N-bound arene rings.

under relatively mild conditions. Such reductions were expanded to a series of substituted anilines, as well as the aziridine PhN(CHPh)$_2$ and ketimine PhN=CPh(Me). In the latter cases of the aziridine and ketimine, reductive ring-opening or imine hydrogenation with concurrent reduction of the nitrogen-bound phenyl ring yielded the products [PhCH$_2$CH(Ph)N(Cy)H$_2$][HB(C$_6$F$_5$)$_3$] and [PhCH(Me)N(Cy)H$_2$][HB(C$_6$F$_5$)$_3$], respectively (Scheme 3.38). These reactions require 4–8 equivalents of dihydrogen with repetitive dihydrogen activation. However, the reductions yield a more basic N-atom, and consequently the borane is sequestered as the ammonium salts. Thus, these cannot be viewed as catalytic reactions.

Computational studies of these aniline reductions showed that the low-temperature activation of dihydrogen is exothermic by 9.7 kcal mol^{-1} and thus is reversible at elevated temperatures. In addition to the B/N FLP, the FLP derived from the Lewis acidic borane and the basic *para*-C atom of aniline is also accessible. Activation of dihydrogen by the latter delivers a proton to the aromatic ring, with a computed barrier of only 8.7 kcal mol^{-1}. This addition irreversibly disrupts the aromaticity and makes subsequent addition of dihydrogen to the nominally enamine fragments thermodynamically downhill, thus facilitating reduction of the arene ring. This latter process to the cyclohexylammonium salts has an activation barrier of *ca.* 20 kcal mol^{-1} and is not reversible (Scheme 3.39).

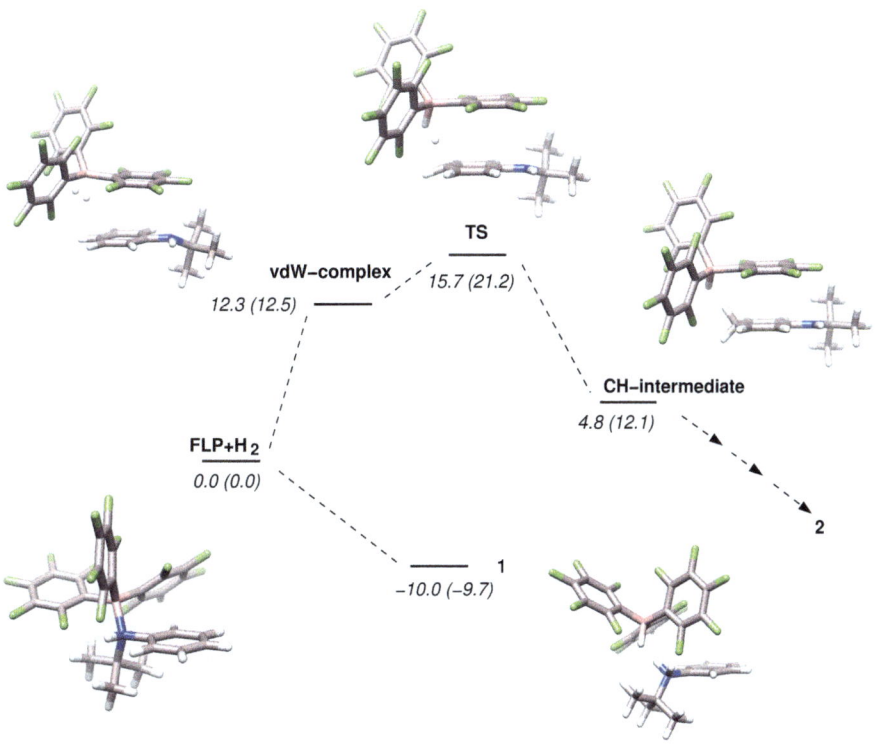

Scheme 3.39 Computed reaction pathway for reduction of aniline to cyclohexylamine. Reproduced from ref. 97 with permission from American Chemical Society, Copyright 2012.

3.12 FLP Reductions of Carbonyl Groups

One functional group that remained a target for FLP reduction was carbonyl moieties. While a computational study suggested rather exotic intramolecular FLPs could affect ketone hydrogenation,[98] experimental studies were delayed due to the thought that the known oxophilicity of boron might preclude effective reduction. In addition, in an early effort to effect carbonyl hydrogenation, toluene solutions of aliphatic ketones in the presence of $B(C_6F_5)_3$ and hydrogen resulted in the stoichiometric formation of the corresponding borinic ester $R_2CHOB(C_6F_5)_2$ at elevated temperatures (Scheme 3.40), concurrent with the release of HC_6F_5.[99] Mechanistically, these reactions proceed *via* FLP dihydrogen activation between the ketone and the borane, transiently forming the corresponding alcohol, which subsequently reacts with the borane to produce the borinic ester (Scheme 3.40). These data seemed to infer that the FLP reduction of carbonyl groups might be problematic. However, a remarkably simple solution overcame this issue.

The Stephan[100] and Ashley[101] groups simultaneously recognized that by simply altering the solvent, the FLP hydrogenation of the carbonyl group was possible. Stephan and coworkers employed diethyl ether or diisopropyl ether as the solvent to affect the

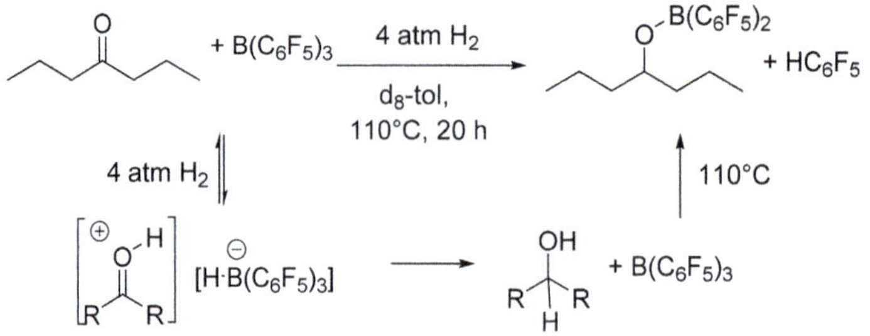

Scheme 3.40 Reactions of ketones with $B(C_6F_5)_3$ in toluene.

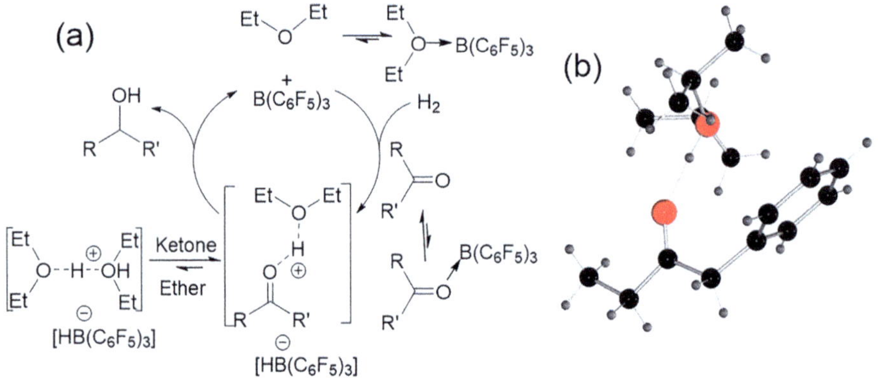

Scheme 3.41 (a) Mechanism of ketone hydrogenation by FLP. (b) POV-ray depiction of the cation $[(i\text{-}Pr_2O)HO=C(CH_2Ph)Et]^+$.

hydrogenation of ketones using a catalytic amount of $B(C_6F_5)_3$ at 70 °C and 5 bar of dihydrogen. Ashley *et al.* employed tetrahydrofuran or 1,4-dioxane as the solvent in a similar fashion, although they employed lower temperatures and pressures, thus requiring longer reaction times. These protocols proved highly successful over a large substrate scope of aliphatic and aromatic ketones and aldehydes.

The dramatic alteration of reactivity upon the use of ethereal solvents was attributed to the ability of these solvents to engage in hydrogen bonding to the transiently protonated ketone. Such interactions were thought to preclude the reaction of the proton with the borane, thereby precluding proto-deborylation of a C_6F_5 group. At the same time, the hydrogen bonding enhances the electrophilicity at the carbonyl carbon favoring hydride delivery from the borohydride anion to give the product alcohol and regenerating free borane and solvent (Scheme 3.41).

This view of the reaction pathway was further supported by the reaction of 1-phenyl-2-butanone, diisopropyl ether, and Jutzi's acid $[(i\text{-}Pr_2O)H][B(C_6F_5)_4]$. This afforded $[(i\text{-}Pr_2O)HO=C(CH_2Ph)Et][B(C_6F_5)_4]$, which was characterized crystallographically (Scheme 3.41). The structural data confirmed the hydrogen-bonding interaction between the carbonyl oxygen atom of the ketone and the protonated ether. This cation is directly analogous to that proposed in the mechanism and was only isolable if the non-coordinating anion $[B(C_6F_5)_4]$ was employed. In a subsequent computational study, Pati[102] provided further support for the role of the ethereal solvent as the basic component of the FLP in the activation of dihydrogen and its participation in hydrogen-bonding interactions with the carbonyl fragment.

In a subsequent study, Stephan and co-workers[103] described a modified protocol for ketone reduction using $B(C_6F_5)_3$ in the presence of molecular sieves in toluene. This simple method also gave hydrogenation of aliphatic, benzyl, and cyclic ketones in high yields, and was easily performed on a gram scale. Further efforts to replace the sieves with α-cyclodextrin (α-CD) also led to ketone reduction, although these reactions were not stereoselective. The mechanism of these reductions was thought to be like that described above, with the heterogeneous materials replacing the ethereal solvent in the stabilization of the transiently protonated ketone. Most conveniently these materials are easily separable and subsequent removal of the borane from the product is achieved by filtration through Celite.

Ashley and coworkers[104] further improved the carbonyl reductions by disproving the misconception that $B(C_6F_5)_3$ is moisture sensitive. They showed that the reduction of ketones including acetone could be performed using $B(C_6F_5)_3$ in 1,4-dioxane without the need for rigorously anhydrous conditions. Similarly, the Soós group[105] exploited their air-stable boranes such as $B(2,6-C_6Cl_2H_3)(2,3,5,6-C_6F_4H)_2$ or $B(2,3,6-C_6Cl_3H_2)$ $(2,3,5,6-C_6F_4H)_2$ to mediate bench-top hydrogenations of a wide array of ketones and aldehydes to the corresponding alcohols. Privalov and coworkers have also probed the details of the mechanism of carbonyl reductions computationally.[106,107]

In a more recent study on carbonyl reductions, the Soós group[108] developed a tandem process in which borane mediated the activation of dihydrogen and alcohols. In this fashion, a ketone in the presence of alcohol is reduced initially to the corresponding acetal and then further reduced to the product ether. Thus, by employing $B(2,6-C_6Cl_2H_3)(2,3,5,6-C_6F_4H)_2$ or $B(2,3,6-C_6Cl_3H_2)(2,3,5,6-C_6F_4H)_2$ as catalysts in THF under dihydrogen, the species $[C_4H_8OH][HBAr_3]$ and $[C_4H_8OH][ROBAr_3]$ are accessible in solution. Protonation of the ketone substrate and subsequent reaction with alcohol generates an equilibrium mixture of acetal and oxocarbenium cation (Scheme 3.42). This latter species can accept hydride from the borohydride anion to give the product ether. The net result is the reduction of a carbonyl species to an ether. It is also notable that the more sterically demanding boranes were effective whereas $B(C_6F_5)_3$ led only to the formation of the acetal product.

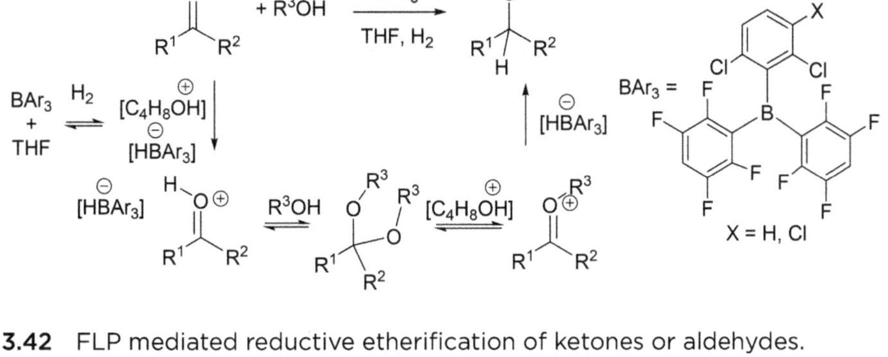

Scheme 3.42 FLP mediated reductive etherification of ketones or aldehydes.

In an interesting exploration of the application of FLP carbonyl reductions for isotopic enrichment, the ability to activate dihydrogen with $C_5H_6Me_4NH$ and $B(C_6F_5)_3$ was employed to reduce the carbonyl species $HC(O)C_6H_4NHBoc$ to the corresponding alcohol. Using D_2 or T_2 gas, this afforded the isotopically labeled products $HDC(OH)C_6H_4NHBoc$ and $HTC(OH)C_6H_4NHBoc$ with 96 and 84% specific enrichment, respectively.[109]

In a unique approach to the reduction of carbonyl groups, Krempner and coworkers[110] exploited what they called 'inverse FLPs'. These are FLP systems derived from comparatively strong bases and relatively weak Lewis acids compared to those dominant in the literature. For example, exploiting $(C_4H_8N)_3PNt$-Bu in the presence of $F_3CC_6H_4$-BBN, these authors affected the reduction of a series of ketones at 75 °C and 100 bar of dihydrogen pressure in 20–40 h with typical yields of *ca.* 90%.

3.13 Implications

The application of the activation of dihydrogen by FLPs to metal-free catalytic reduction began with imines and evolved rapidly. Catalyst variation and modifications offered enhanced activity, air stability, and consequently ease of use in the lab. Further variations of the catalysts have offered considerable structural variety and served to affirm the generality of the concept of 'frustration' in the development of metal-free hydrogenation catalysts.

It is, however, the broadening of the substrate scope that has perhaps uncovered the greatest potential for these metal-free FLP reductions. Extensions to N-heterocycles, olefins, alkynes, aromatic species, and carbonyl moieties have been developed and these firmly establish metal-free hydrogenations as a viable alternative to metal-based hydrogenations.

It is perhaps informative to consider the parallels between the advent of transition metal and FLP catalysts, even though these occurred some 50 years apart. In both cases, after the early findings that select species could act as catalysts, efforts were undertaken to improve the activity. While initial efforts made some improvements, strategies for catalyst optimization, tuning, and design emerged as we began to understand these respective systems more clearly. In both transition metal and FLP chemistry, efforts to broadening the substrate scope followed.

Having discussed these advances in this chapter, we are now primed for the next stage of development. Following the evolution of organometallic hydrogenation catalysts, we are now positioned to ask perhaps the most challenging question regarding the possibility of stereo-selective reductions. Thus: can we modify the FLP catalyst for metal-free asymmetric reductions? This aspect is addressed in the next chapter.

References

1. C. Walling and L. Bollyky, *J. Am. Chem. Soc.*, 1961, **83**, 2968–2969.
2. C. Walling and L. Bollyky, *J. Am. Chem. Soc.*, 1964, **86**, 3750–3752.
3. J. Spielmann, F. Buch and S. Harder, *Angew. Chem., Int. Ed.*, 2008, **47**, 9434–9438.
4. A. Berkessel, T. J. S. Schubert and T. N. Mueller, *J. Am. Chem. Soc.*, 2002, **124**, 8693–8698.
5. R. Noyori and T. Ohkuma, *Angew. Chem., Int. Ed.*, 2001, **40**, 40–73.

6. T. Ohkuma, N. Utsumi, K. Tsutsumi, K. Murata, C. Sandoval and R. Noyori, *J. Am. Chem. Soc.*, 2006, **128**, 8724–8725.
7. T. Kimura, T. Takahashi, M. Nishiura and K. Yamamura, *Org. Lett.*, 2006, **8**, 3137–3139.
8. M. Siskin, *J. Am. Chem. Soc.*, 1974, **96**, 3640–3641.
9. J. Wristers, *J. Am. Chem. Soc.*, 1975, **97**, 4312–4316.
10. M. Yalpani, R. Köster and M. W. Haenel, *Erdöl & Kohle Erdgas Petrochemie*, 1990, **43**, 344–347.
11. M. Yalpani, T. Lunow and R. Köster, *Chem. Ber.*, 1989, **122**, 687–693.
12. M. W. Haenel, J. Narangerel, U. B. Richter and A. Rufińska, *Angew. Chem., Int. Ed.*, 2006, **45**, 1061–1066.
13. P. A. Chase, G. C. Welch, T. Jurca and D. W. Stephan, *Angew. Chem., Int. Ed.*, 2007, **46**, 8050–8053.
14. P. A. Chase, T. Jurca and D. W. Stephan, *Chem. Commun.*, 2008, 1701–1703.
15. G. C. Welch, L. Cabrera, P. A. Chase, E. Hollink, J. D. Masuda, P. R. Wei and D. W. Stephan, *Dalton Trans.*, 2007, 3407–3414.
16. P. Spies, S. Schwendemann, S. Lange, G. Kehr, R. Fröhlich and G. Erker, *Angew. Chem., Int. Ed.*, 2008, **47**, 7543–7546.
17. S. M. Whittemore, G. Edvenson, D. M. Camaioni, A. Karkamkar, D. Neiner, K. Parab and T. Autrey, *Catal. Today*, 2015, **251**, 28–33.
18. D. J. Chen and J. Klankermayer, *Chem. Commun.*, 2008, 2130–2131.
19. H. D. Wang, R. Fröhlich, G. Kehr and G. Erker, *Chem. Commun.*, 2008, 5966–5968.
20. C. Jiang, O. Blacque and H. Berke, *Chem. Commun.*, 2009, 5518–5520.
21. V. Sumerin, F. Schulz, M. Atsumi, C. Wang, M. Nieger, M. Leskelä, T. Repo, P. Pyykkö and B. Rieger, *J. Am. Chem. Soc.*, 2008, **130**, 14117–14118.
22. K. V. Axenov, G. Kehr, R. Fröhlich and G. Erker, *J. Am. Chem. Soc.*, 2009, **131**, 3454–3455.
23. S. Tussing, L. Greb, S. Tamke, B. Schirmer, C. Muhle-Goll, B. Luy and J. Paradies, *Chem. - Eur. J.*, 2015, **21**, 8056–8059.
24. S. Tussing, K. Kaupmees and J. Paradies, *Chem. - Eur. J.*, 2016, **22**, 7422–7426.
25. L. Greb, S. Tussing, B. Schirmer, P. Ona-Burgos, K. Kaupmees, M. Lokov, I. Leito, S. Grimme and J. Paradies, *Chem. Sci.*, 2013, **4**, 2788–2796.
26. S. Tussing and J. Paradies, *Dalton Trans.*, 2016, **45**, 6124–6128.
27. G. Wang, C. Chen, T. Q. Du and W. H. Zhong, *Adv. Synth. Catal.*, 2014, **356**, 1747–1752.
28. L. Greb, P. Oña-Burgos, A. Kubas, F. C. Falk, F. Breher, K. Fink and J. Paradies, *Dalton Trans.*, 2012, **41**, 9056–9060.
29. D. W. Stephan, S. Greenberg, T. W. Graham, P. Chase, J. J. Hastie, S. J. Geier, J. M. Farrell, C. C. Brown, Z. M. Heiden, G. C. Welch and M. Ullrich, *Inorg. Chem.*, 2011, **50**, 12338–12348.
30. G. Eros, H. Mehdi, I. Papai, T. A. Rokob, P. Kiraly, G. Tarkanyi and T. Soós, *Angew. Chem., Int. Ed.*, 2010, **49**, 6559–6563.
31. D. J. Scott, M. J. Fuchter and A. E. Ashley, *Angew. Chem., Int. Ed.*, 2014, **53**, 10218–10222.
32. J. S. Reddy, B. H. Xu, T. Mahdi, R. Fröhlich, G. Kehr, D. W. Stephan and G. Erker, *Organometallics*, 2012, **31**, 5638–5649.
33. G. Eros, K. Nagy, H. Mehdi, I. Papai, P. Nagy, P. Kiraly, G. Tarkanyi and T. Soós, *Chem. - Eur. J.*, 2012, **18**, 574–585.
34. E. Dorko, M. Szabo, B. Kotai, I. Papai, A. Domjan and T. Soós, *Angew. Chem., Int. Ed.*, 2017, **56**, 9512–9516.
35. E. Dorko, B. Kotai, T. Foldes, A. Gyomore, I. Papai and T. Soós, *J. Organomet. Chem.*, 2017, **847**, 258–262.
36. Y. Hoshimoto, T. Kinoshita, S. Hazra, M. Ohashi and S. Ogoshi, *J. Am. Chem. Soc.*, 2018, **140**, 7292–7300.
37. X. Zhu and H. Du, *Org. Lett.*, 2015, **17**, 3106–3109.
38. S. Schwendemann, T. A. Tumay, K. V. Axenov, I. Peuser, G. Kehr, R. Fröhlich and G. Erker, *Organometallics*, 2010, **29**, 1067–1069.
39. Z. Pan, H. Wang, F. Ling, L. Xiao, D. Song and W. Zhong, *Synth. Commun.*, 2019, **49**, 522–528.
40. S. Mummadi, D. K. Unruh, J. Zhao, S. Li and C. Krempner, *J. Am. Chem. Soc.*, 2016, **138**, 3286–3289.
41. J. Mohr and M. Oestreich, *Angew. Chem., Int. Ed.*, 2014, **53**, 13278–13281.
42. J. Mohr, D. Porwal, I. Chatterjee and M. Oestreich, *Chem. - Eur. J.*, 2015, **21**, 17583–17586.
43. N. A. Sitte, M. Bursch, S. Grimme and J. Paradies, *J. Am. Chem. Soc.*, 2019, **141**, 159–162.
44. L. Koring, A. Sitte Nikolai, M. Bursch, S. Grimme and J. Paradies, *Chem. - Eur. J.*, 2021, DOI: 10.1002/chem.202100041.
45. H. Zhu, Z.-W. Qu and S. Grimme, *Eur. J. Org. Chem.*, 2019, **2019**, 4609–4612.
46. S. J. Geier, P. A. Chase and D. W. Stephan, *Chem. Commun.*, 2010, **46**, 4884–4886.
47. T. Mahdi, J. N. del Castillo and D. W. Stephan, *Organometallics*, 2013, **32**, 1971–1978.
48. J. Y. Zhao, G. Q. Wang and S. H. Li, *Dalton Trans.*, 2015, **44**, 9200–9208.

49. J. M. Farrell, Z. M. Heiden and D. W. Stephan, *Organometallics*, 2011, **30**, 4497–4500.
50. G. Li, Y. B. Liu and H. F. Du, *Org. Biomol. Chem.*, 2015, **13**, 2875–2878.
51. Y. B. Liu and H. F. Du, *J. Am. Chem. Soc.*, 2013, **135**, 12968–12971.
52. W. Wang, X. Q. Feng and H. F. Du, *Org. Biomol. Chem.*, 2016, **14**, 6683–6686.
53. W. Wang, W. Meng and H. Du, *Dalton Trans.*, 2015, **45**, 5945–5948.
54. S. Wei, X. Feng and H. Du, *Org. Biomol. Chem.*, 2016, **14**, 8026–8029.
55. Z. B. Jian, S. Krupski, K. Skoch, G. Kehr, C. G. Daniliuc, I. Cisarova, P. Stepnicka and G. Erker, *Organometallics*, 2017, **36**, 2940–2946.
56. L. M. Elmer, G. Kehr, C. G. Daniliuc, M. Siedow, H. Eckert, M. Tesch, A. Studer, K. Williams, T. H. Warren and G. Erker, *Chem. - Eur. J.*, 2017, **23**, 6056–6068.
57. T. Özgün, K. Y. Ye, C. G. Daniliuc, B. Wibbeling, L. Liu, S. Grimme, G. Kehr and G. Erker, *Chem. - Eur. J.*, 2016, **22**, 5988–5995.
58. A. Feldmann, G. Kehr, C. G. Daniliuc, C. Muck-Lichtenfeld and G. Erker, *Chem. - Eur. J.*, 2015, **21**, 12456–12464.
59. C. Rosorius, J. Moricke, B. Wibbeling, A. C. McQuilken, T. H. Warren, C. G. Daniliuc, G. Kehr and G. Erker, *Chem. - Eur. J.*, 2016, **22**, 1103–1113.
60. L. A. Korte, S. Blomeyer, S. Heidemeyer, A. Mix, B. Neumann and N. W. Mitzel, *Chem. Commun.*, 2016, **52**, 9949–9952.
61. L. A. Korte, S. Blomeyer, S. Heidemeyer, J. H. Nissen, A. Mix, B. Neumann, H. G. Stammler and N. W. Mitzel, *Dalton Trans.*, 2016, **45**, 17319–17328.
62. C. Woelke, C. G. Daniliuc, G. Kehr and G. Erker, *J. Organomet. Chem.*, 2019, **899**, 120879.
63. X. Zhao, T. M. Gilbert and D. W. Stephan, *Chem. - Eur. J.*, 2010, **16**, 10304–10308.
64. A. Schnurr, H. Vitze, M. Bolte, H. W. Lerner and M. Wagner, *Organometallics*, 2010, **29**, 6012–6019.
65. A. Schnurr, M. Bolte, H.-W. Lerner and M. Wagner, *Eur. J. Inorg. Chem.*, 2012, **2012**, 112–120.
66. F. Bertini, V. Lyaskoyskyy, B. J. J. Timmer, F. J. J. de Kanter, M. Lutz, A. W. Ehlers, J. C. Slootweg and K. Lammertsma, *J. Am. Chem. Soc.*, 2012, **134**, 201–204.
67. K. Samigullin, I. Georg, M. Bolte, H. W. Lerner and M. Wagner, *Chem. - Eur. J.*, 2016, **22**, 3478–3484.
68. Z. B. Jian, G. Kehr, C. G. Daniliuc, B. Wibbeling and G. Erker, *Dalton Trans.*, 2017, **46**, 11715–11721.
69. J. G. Yu, G. Kehr, C. G. Daniliuc, C. Bannwarth, S. Grimme and G. Erker, *Org. Biomol. Chem.*, 2015, **13**, 5783–5792.
70. Y. Wang, Z. H. Li and H. Wang, *RSC Adv.*, 2018, **8**, 26271–26276.
71. J. J. Cabrera-Trujillo and I. Fernandez, *Chem. - Eur. J.*, 2018, **24**, 17823–17831.
72. D. Yepes, P. Jaque and I. Fernandez, *Chem. - Eur. J.*, 2018, **24**, 8833–8840.
73. G. Lu, H. Li, L. Zhao, F. Huang and Z. Wang, *Inorg. Chem.*, 2010, **49**, 295–301.
74. S. Tamke, C. G. Daniliuc and J. Paradies, *Org. Biomol. Chem.*, 2014, **12**, 9139–9144.
75. L. Greb, P. Oña-Burgos, B. Schirmer, S. Grimme, D. W. Stephan and J. Paradies, *Angew. Chem., Int. Ed.*, 2012, **51**, 10164–10168.
76. P. Jutzi, C. Müller, A. Stammler and H. G. Stammler, *Organometallics*, 2000, **19**, 1442–1444.
77. L. J. Hounjet, C. Bannwarth, C. N. Garon, C. B. Caputo, S. Grimme and D. W. Stephan, *Angew. Chem., Int. Ed.*, 2013, **52**, 7492–7495.
78. S. M. Whittemore and T. Autrey, *Isr. J. Chem.*, 2015, **55**, 196–201.
79. B. Inés, D. Palomas, S. Holle, S. Steinberg, J. A. Nicasio and M. Alcarazo, *Angew. Chem., Int. Ed.*, 2012, **51**, 12367–12369.
80. J. A. Nicasio, S. Steinberg, B. Inés and M. Alcarazo, *Chem. - Eur. J.*, 2013, **19**, 11016–11020.
81. L. Greb, C. G. Daniliuc, K. Bergander and J. Paradies, *Angew. Chem., Int. Ed.*, 2013, **52**, 5876–5879.
82. V. Morozova, P. Mayer and G. Berionni, *Angew. Chem., Int. Ed.*, 2015, **54**, 14508–14512.
83. N. S. Lambic, R. D. Sommer and E. A. Ison, *J. Am. Chem. Soc.*, 2016, **138**, 4832–4842.
84. N. S. Lambic, R. D. Sommer and E. A. Ison, *ACS Catal.*, 2017, **7**, 1170–1180.
85. N. S. Lambic, R. D. Sommer and E. A. Ison, *Dalton Trans.*, 2020, **49**, 6127–6134.
86. G. I. Nikonov, S. F. Vyboishchikov and O. G. Shirobokov, *J. Am. Chem. Soc.*, 2012, **134**, 5488–5491.
87. Y. Wang, W. Chen, Z. Lu, Z. H. Li and H. Wang, *Angew. Chem., Int. Ed.*, 2013, **52**, 7496–7499.
88. I. Khan, M. Manzotti, G. J. Tizzard, S. J. Coles, R. L. Melen and L. C. Morrill, *ACS Catal.*, 2017, **7**, 7748–7752.
89. B. H. Xu, G. Kehr, R. Frohlich, B. Wibbeling, B. Schirmer, S. Grimme and G. Erker, *Angew. Chem., Int. Ed.*, 2011, **50**, 7183–7186.
90. K. Chernichenko, Á. Madarász, I. Pápai, M. Nieger, M. Leskelä and T. Repo, *Nat. Chem.*, 2013, **5**, 718–723.

91. K. Chernichenko, B. Kotai, M. Nieger, S. Heikkinen, I. Papai and T. Repo, *Dalton Trans.*, 2017, **46**, 2263–2269.
92. Y. B. Liu, L. R. Hu, H. Chen and H. F. Du, *Chem. - Eur. J.*, 2015, **21**, 3495–3501.
93. U. Gellrich, *Angew. Chem., Int. Ed.*, 2018, **57**, 4779–4782.
94. F. Wech, M. Hasenbeck and U. Gellrich, *Chem. - Eur. J.*, 2020, **26**, 13445–13450.
95. M. Ghara, S. Pan and P. K. Chattaraj, *Phys. Chem. Chem. Phys.*, 2019, **21**, 21267–21277.
96. Y. Segawa and D. W. Stephan, *Chem. Commun.*, 2012, **48**, 11963–11965.
97. T. Mahdi, Z. M. Heiden, S. Grimme and D. W. Stephan, *J. Am. Chem. Soc.*, 2012, **134**, 4088–4091.
98. H. Li, L. Zhao, G. Lu, F. Huang and Z.-X. Wang, *Dalton Trans.*, 2010, **39**, 5519–5526.
99. L. E. Longobardi, C. Tang and D. W. Stephan, *Dalton Trans.*, 2014, **43**, 15723–15726.
100. T. Mahdi and D. W. Stephan, *J. Am. Chem. Soc.*, 2014, **136**, 15809–15812.
101. D. J. Scott, M. J. Fuchter and A. E. Ashley, *J. Am. Chem. Soc.*, 2014, **136**, 15813–15816.
102. S. Das and S. K. Pati, *Chem. - Eur. J.*, 2017, **23**, 1078.
103. T. Mahdi and D. W. Stephan, *Angew. Chem., Int. Ed.*, 2015, **54**, 8511–8514.
104. D. J. Scott, T. R. Simmons, E. J. Lawrence, G. G. Wildgoose, M. J. Fuchter and A. E. Ashley, *ACS Catal.*, 2015, **5**, 5540–5544.
105. Á. Gyömöre, M. Bakos, T. Földes, I. Pápai, A. Domján and T. Soós, *ACS Catal.*, 2015, **5**, 5366–5372.
106. M. Heshmat and T. Privalov, *Chem. - Eur. J.*, 2017, **23**, 9098–9113.
107. M. Heshmat and T. Privalov, *J. Phys. Chem. B*, 2018, **122**, 8952–8962.
108. M. Bakos, A. Gyomore, A. Domjan and T. Soós, *Angew. Chem., Int. Ed.*, 2017, **56**, 5217–5221.
109. A. Marek and M. H. F. Pedersen, *Tetrahedron*, 2015, **71**, 917–921.
110. S. Mummadi, A. Brar, G. Wang, D. Kenefake, R. Diaz, D. K. Unruh, S. Li and C. Krempner, *Chem. - Eur. J.*, 2018, **24**, 16526–16531.

4 Borane-based Asymmetric FLP Hydrogenations

4.1 Chapter Overview

Having established the ability of FLPs to activate dihydrogen and mediate the hydrogenation of a variety of organic substrates, the field was poised for the next step in its evolution. Paralleling the development of transition metal-based hydrogenation catalysis, stereoselective asymmetric FLP hydrogenation was targeted.

In this chapter, we discuss the early literature that established the potential of such selectivity in FLP reductions. Initial efforts employing chirality within substrates to encourage diastereoselective reductions are considered. Following that, studies that introduce chirality into the catalyst to access enantioselective reductions are discussed. Despite poor enantioselectivity in the initial catalysis, the evolution of chiral FLP catalysts proved to be rapid and several highly selective borane-based catalyst systems have emerged, In addition, the application of such systems to a growing range of prochiral substrates has continued to develop. The chapter is completed with a consideration of the key design features that have emerged.

4.2 Diastereoselective Reactions

As we have described in Chapter 2 (see Section 2.4), there is a strong relationship between the borane-mediated hydrosilylation of ketones and the FLP activation of dihydrogen. Thus, in the context of asymmetric reductions, it is important to note the 2009 studies from Hog and Oestreich[1] in which the authors described the hydrosilylation of ketones with an optically pure silane. In these experiments, a series of ketones are reacted with (R)-H(i-Pr)SiCH$_2$(CH$_2$)$_2$C$_6$H$_4$ in the presence of 5 mol% B(C$_6$F$_5$)$_3$. In the case of Ph(Me)CO, the resulting product is the silyl ether obtained in >90% yield. Moreover, while the chirality at silicon has been inverted (>97%), the product

A Primer in Frustrated Lewis Pair Hydrogenation: Concepts to Applications
By Douglas W. Stephan
© Douglas W. Stephan 2022
Published by the Royal Society of Chemistry, www.rsc.org

is a diastereomeric 74:26 mixture of the $^{Si}R,R$- and $^{Si}R,S$-isomers (Scheme 4.1). These authors showed similar results for several aryl-methyl ketones, typically affording an 80:20 mixture of the diastereomers. In contrast, efforts to catalyze the hydrosilylation of methyl benzylimines, Ph(Me)C=NR (R = Bz, Ph) were unsuccessful. The authors attributed this latter result to the stability of the imine–borane adduct. Nonetheless, the use of a stoichiometric amount of borane resulted in a racemic reduction to the corresponding amine.

Relating to this latter observation, Massey and Park had described the ability of $B(C_6F_5)_3$ to form classical adducts with amines.[2] However, it must also be noted that Basset and co-workers[3] reported that the stoichiometric reaction of $B(C_6F_5)_3$ and Et_2NPh did not give an adduct but rather an equilibrium mixture of free borane, amine, the salt $[Et_2NHPh][HB(C_6F_5)_3]$ and the zwitterion $[EtPhN=CHCH_2B(C_6F_5)_3]$, inferring the ability of the borane to abstract a benzylic hydride. Similarly, the groups of Resconi[4] and Rieger[5] observed analogous products for reactions of Et_2NH and i-Pr_2NH with $B(C_6F_5)_3$, respectively. In subsequent work, we related these observations to those of Oestreich, as the combination of the optically pure bis((R)-1-phenylethyl)amine, with 1 mol% $B(C_6F_5)_3$ was shown to give an equilibrium mixture of diastereomers of the amine (Scheme 4.2).[6] This observation demonstrates the ability of $B(C_6F_5)_3$ to reversibly abstract a hydride from the chiral benzylic positions of the amine, prompting racemization. While this observation was exploited to effect transfer hydrogenation (see Section 6.3), for FLP-based asymmetric reductions, these observations demonstrate that catalyst modification is not the only consideration for the enantioselective hydrogenation. One must also assess the potential of the Lewis acid to mediate the racemization of the chiral product.

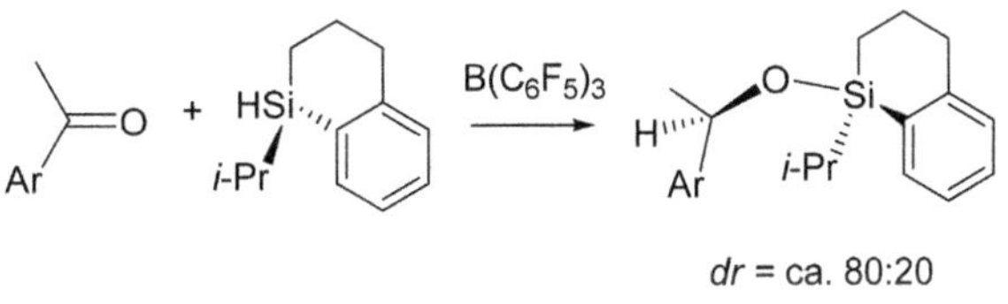

dr = ca. 80:20

Scheme 4.1　Diastereoselective hydrosilylation of the methyl-aryl ketone.

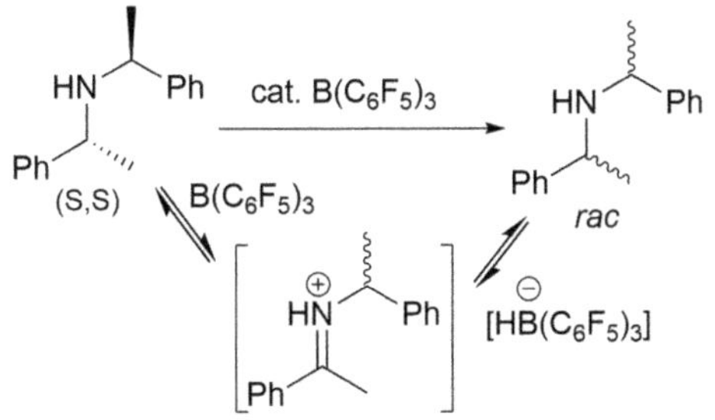

Scheme 4.2　Borane mediated racemization of bis(1-phenylethyl)amine.

We also noted that the Soós report[7] of the hydrogenation of D-(+)-carvone mediated by 20 mol% of the FLP derived from MesB(C$_6$F$_5$)$_2$ and DABCO led to the reduction of the C=C double bond, leaving the carbonyl untouched (see Section 3.9). Interestingly the product was found to be a mixture of the stereoisomers, enriched in *trans*-isomer (*trans*:*cis* = 4.3:1). This observation infers that the stereogenic centers tolerate the presence of this less Lewis acidic borane without epimerization. Perhaps more importantly, it demonstrates the influence of chiral centers in the substrate molecules on the stereochemistry of delivery of dihydrogen to the unsaturated bond.

In our work, B(C$_6$F$_5$)$_3$ was shown to reduce optically pure ketimines in a diastereoselective manner.[8] In the case of imines in which the chiral center is adjacent to the nitrogen center, reductions performed using 5 mol% B(C$_6$F$_5$)$_3$ under 5 bar of dihydrogen at 80 °C gave quantitative reduction after 24–48 h. The resulting amines were isolated with diastereomeric excesses ranging from 0% to 65%. Camphor or menthone-derived imines were similarly reduced, although they were performed at 120 °C and required 120 h for complete conversion. In these cases, diastereomeric excesses were much higher, ranging from 96% to 99%. In these latter cases, the proposed intermediate iminium hydridoborate is thought to provide a sterically encumbered, chiral environment about the iminium carbon, thus generating a preferred face for hydride delivery (Scheme 4.3).

Erker and coworkers[9] studied the reductions of chiral imine-derived ferrocenes. The chiral species R,R-(C$_5$H$_3$(NCPhR)CH(NMe$_2$)CH$_2$CH(Me)C$_5$H$_4$)Fe (R = CF$_3$, Me) were reduced in the presence of 10 mol% C$_6$H$_{13}$B(C$_6$F$_5$)$_2$, giving a 5:1 mixture of the diastereomers. The species R,R-(C$_5$H$_3$ (NCPhR)CH(PMes$_2$)CH$_2$CH(Me)C$_5$H$_4$)Fe was also reduced under similar condition. However, the product underwent P–C bond cleavage affording a 2.2:1 mixture of the diastereomeric products (C$_5$H$_3$(NHCHPhR)CH$_2$CH$_2$CH(Me)C$_5$H$_4$)Fe (Scheme 4.4).

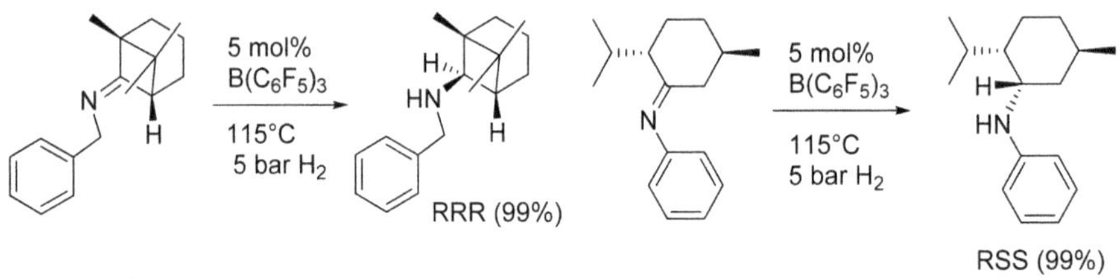

Scheme 4.3 Diastereoselective reduction of Camphor or menthone-derived imines.

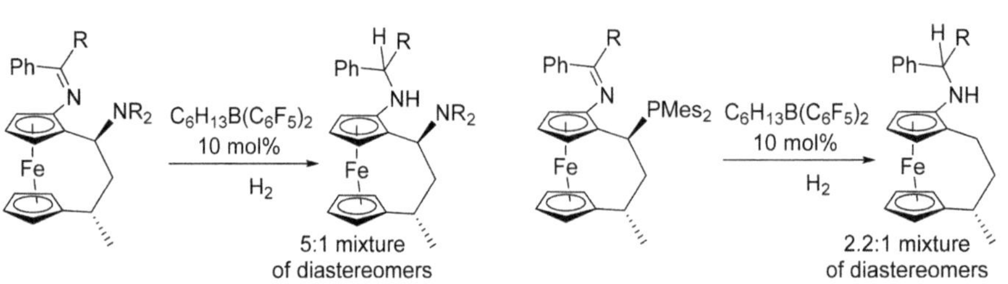

Scheme 4.4 Reductions of chiral imine-derived ferrocenes.

4.3 Considerations for Enantioselective Catalysts

Our initial efforts towards enantioselective FLP reductions began with the idea of using chiral components that were readily available.[10] Certainly, chiral bidentate phosphines have been widely employed in the development of transition metal-based asymmetric catalysts. Thus, in initial trials, FLPs were generated employing the optically pure phosphines, (*R*)-binap, (*S,S*)-chiraphos, or (*S,S*)-diop in combination with the Lewis acid $B(C_6F_5)_3$. Using 20 mol% catalyst and 4 bar of dihydrogen, the imine PhC(Me)=NPh was reduced at temperatures from 50–100 °C. The FLPs derived from (*R*)-binap or (*S,S*)-chiraphos gave the racemic amine, while the FLP derived from (*S,S*)-diop afforded an enantiomeric excess of 25% (Scheme 4.5).

Putting these observations in the light of what was known about the mechanism, it is clear the initial logic was flawed. Imine reduction is known to involve consecutive addition of proton and hydride. Thus, the only way in which a chiral base can impact the stereoselectivity in the hydride delivery is if the reduction occurs in a concerted fashion. While this may be the case for intramolecular FLPs, this seems unlikely for intermolecular FLPs. In the latter case, proton delivery to the nitrogen of the imine generates the iminium cation that is sufficiently acidic to accept the hydride. Thus, any base–iminium interaction would only serve to diminish the acidity of the iminium carbon and slow the reduction. Based on this view, efforts to develop enantioselective FLP catalysts have focused on the generation of chiral Lewis acids and chiral intramolecular FLPs.

In a related recent effort, Paradies and coworkers[11] have reported the activation of dihydrogen by the FLP derived from chiral amidinates with $B(2,4,6-F_3C_6H_2)_3$ and $B(2,6-F_2C_6H_3)_3$. These chiral systems affect the hydrogenation of electron-deficient olefins in high yields. However, no asymmetric induction was observed for prochiral substrates (Scheme 4.6). This was attributed to the conformational flexibility of the hydrogen bond donor.

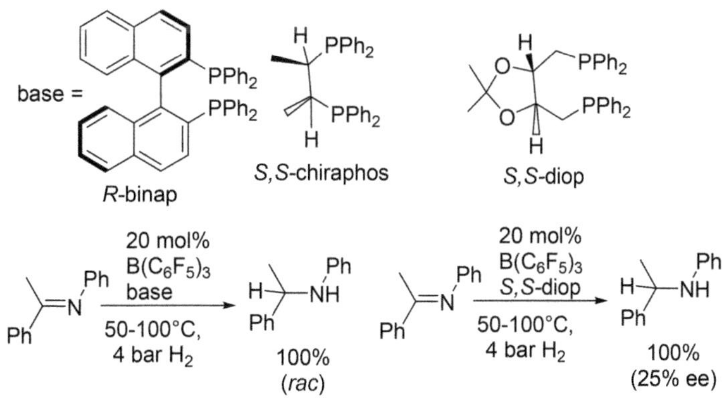

Scheme 4.5 Use of chiral bis-phosphines in FLP reduction of a ketimine.

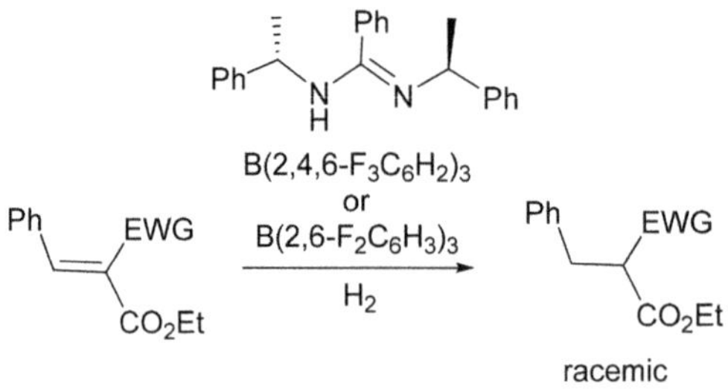

Scheme 4.6   Use of a chiral amidinate in the reduction of a prochiral olefin.

4.4   Chiral Boranes and Enantioselective Reduction

Shortly after the report of metal-free hydrogenations of imines by $Mes_2P(C_6F_4)B(C_6F_5)_2$, it became clear that imine reductions could be mediated by catalytic amounts of borane, as the substrate would act as the basic component of the FLP (See Section 3.3). Almost concurrently with our report, Chen and Klankermayer[12] described related results with several imines. In addition, these authors saw the potential for asymmetric reductions. To that end, they prepared the chiral borane $(C_6H_8(CMe_2)Me)B(C_6F_5)_2$ derived from the hydroboration of α-pinene by Piers' borane.[13] Using 10 mol% of this chiral borane, $Ph(Me)C=NPh$ was quantitatively reduced at 65 °C and 20 bar of dihydrogen in toluene. The product was obtained in 13% enantiomeric excess (Scheme 4.7). While this selectively is far from impressive, this was the first metal-free enantioselective reduction and foreshadowed greater success to come.

In a subsequent study, the Klankermayer group expanded on the initial report by preparing a series of related chiral boranes derived from camphor derivatives.[14] Hydroboration of the bicycloheptene derived from camphor gave a mixture of diastereomers. Subsequent reaction with t-Bu$_3$P and dihydrogen converted these boranes to the corresponding phosphonium hydridoborate salts. Under controlled conditions, the varying reactivity of the diastereomeric boranes with phosphine and dihydrogen allowed the isolation of the diastereomerically pure salts. Using the $[t$-Bu$_3$PH] salt of the (C_6F_5)-(($^1R,^2R,^3R,^4S$)-4,7,7-trimethyl-3-phenylbicyclo[2.2.1]heptan-2-yl)hydroborate anion (Scheme 4.6), several prochiral imines were hydrogenated in high yields using 25 bar of dihydrogen at 65 °C in 15 h. The use of this chiral catalyst gave enantiomeric excesses ranging from 74–83%. This dramatic improvement over the initial results prompted Klankermayer to apply this catalyst in the asymmetric hydrosilylation of imines.[15]

In a 2012 continuation of these efforts, the Klankermayer group described a modification of the above camphor-based chiral borane with the preparation of a closely related zwitterion, $C_6H_7C(Me)_2Me(C_6H_4PHt$-Bu$_2)(BH(C_6F_5)_2)$, (Scheme 4.8).[16] This species was employed to mediate the hydrogenation of imines with enantioselectivities of up to 76% enantiomeric excess. This catalyst also proved to be highly robust and thus amenable to recycling through five catalytic runs before the yields of products were impacted.

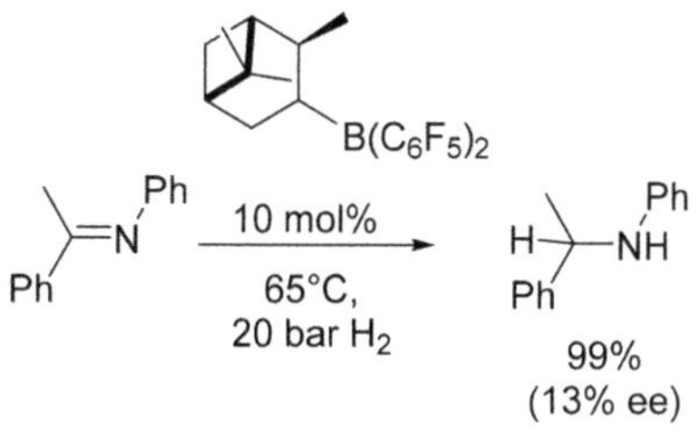

Scheme 4.7 FLP-based enantioselective hydrogenation of imine.

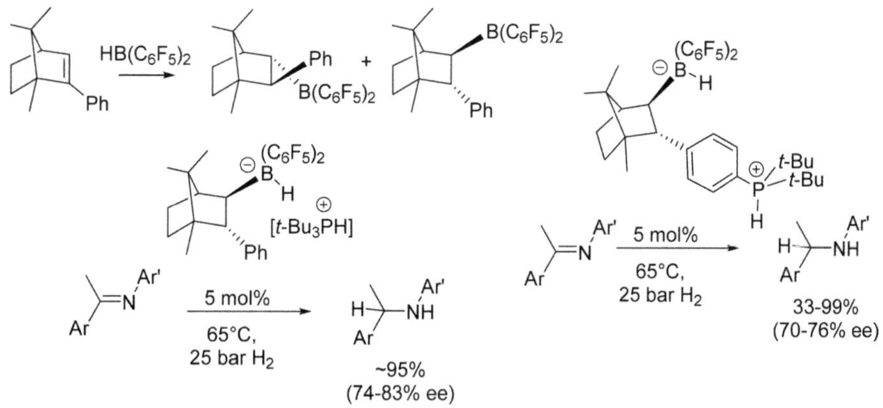

Scheme 4.8 FLP-based enantioselective hydrogenation of imines using camphor-based borane.

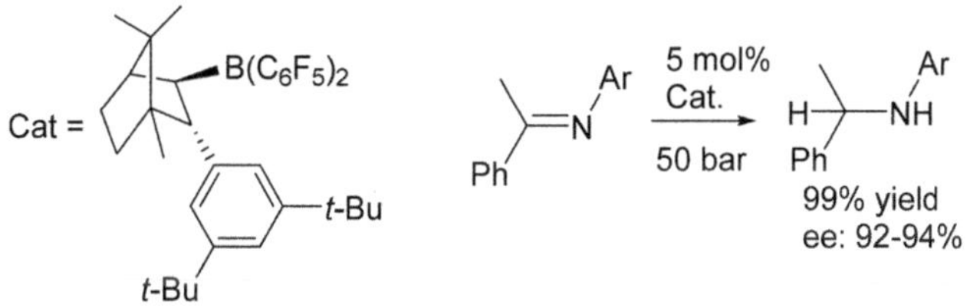

Scheme 4.9 Enantioselective FLP reductions of imines.

In a very recent paper, Papai *et al.*[17] have employed computational chemistry to probe the stereoselectivity-determining hydride transfer process for a series of camphor-derived chiral boranes. These authors predicted that compounds of the form $C_6H_6MeC(Me)(3,5-R_2C_6H_3)(B(C_6F_5)_2)$ (R = F, CF_3, *t*-Bu) would provide high enantioselectivity. Indeed, the compound with R = *t*-Bu was prepared, and using 5 mol% of this species as the catalyst, it mediated the hydrogenation of a series of N-aryl imines in 99% yield after 24 h using 50 bar pressure with enantiomeric excesses of 92–94% (Scheme 4.9).

4.5   Chiral Intramolecular FLPs

Targeting a chiral intramolecular FLP, the Erker group[18] attempted the hydroboration of an enamine derived from camphor. While the product was not isolated, in the presence of dihydrogen, this mixture afforded the salt $(C_6H_7MeC(Me)_2)(=NC_5H_{10})(BH(C_6F_5)_2)$. In related work, Repo group[19] derived a chiral FLP *via* the hydroboration of a camphor enamine obtaining the corresponding zwitterionic iminium-hydridoborate $(C_6H_7C(Me)_2)$ $(=NC_5H_{10})(BH(C_6F_5)_2)$ (Scheme 4.10). Subsequent activation of dihydrogen gave a mixture of isomers of the ammonium-hydridoborate. On the other hand, the corresponding dimethyl amino iminium-hydridoborate derivative reacts with dihydrogen, leading to B–C bond cleavage affording the bornylamine–$HB(C_6F_5)_2$ adduct (Scheme 4.10).

Repo and coworkers developed a synthetic approach to a series of chiral B/N FLPs[20] incorporating amines linked *via* a benzylic fragment in which a $B(C_6F_5)_2$ unit was installed on the arene ring. Judicious choice of the amine allowed access to chiral intramolecular B/N FLP systems. However, despite the use of optically pure amines, systems with hydrogen atoms *beta* to the nitrogen gave a mixture of diastereomers. This was attributed to a reversible intramolecular hydride abstraction affording an avenue to racemization (Scheme 4.11). This observation is akin to the previously described epimerization of chiral amines (see Section 4.2).

With the above lessons, systems that excluded *beta*-hydrogen atoms were targeted to provide stable and enantiomerically pure chiral B/N FLPs.[20] Thus, use of 2,2,4,7 -tetramethyl-1,2,3,4-tetrahydroquinoline or (2R,4aR,9aS)-2-isopropyl-4a-methyl-9a-phenyl-2,3,4,4a,9,9a-hexahydro-1H-carbazole were incorporated into chiral

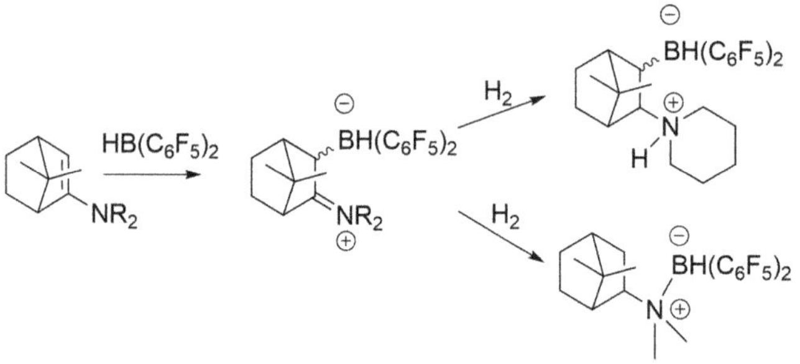

Scheme 4.10   Reactions of iminium-hydridoborate with dihydrogen.

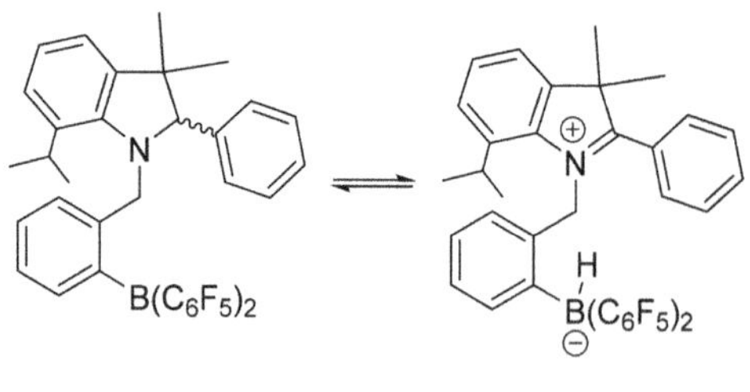

Scheme 4.11   Racemization of an intramolecular chiral B/N FLP.

B/N FLPs (Scheme 4.12). These species were evaluated in the asymmetric reduction of imines and substituted quinolines. While the former B/N FLP effected a quantitative reduction of these substrates, the observed enantiomeric excesses were relatively low, with a maximum of 35%. The chiral carbazole-derived FLP proved to be both less reactive and less selective affording imine reduction as the conversions ranged from 35–70% with enantiomeric excesses of only 8–17%.

Undaunted, Repo and co-workers[21] extended their efforts, targeting another chiral intramolecular B/N FLP. In a new system, a chiral binaphthyl backbone was used to link the borane and amine sites, thus obtaining the catalyst S-$(C_{20}H_{12})NMe(i$-Pr$)$ $(B(C_6F_5)_2)$ (Scheme 4.13). This species, dubbed "molecular tweezers" proved effective in the asymmetric hydrogenation of unhindered N-alkylimines, with enantiomeric excesses ranging from 32% to 83%. These authors also demonstrated that enamines could also be effectively reduced with enantiomeric excesses varied from 47–99%. Computational studies probed the nature of the relevant transition states, revealing a relatively low barrier to dihydrogen splitting by the N/B FLP (14.7 kcal mol^{-1}).

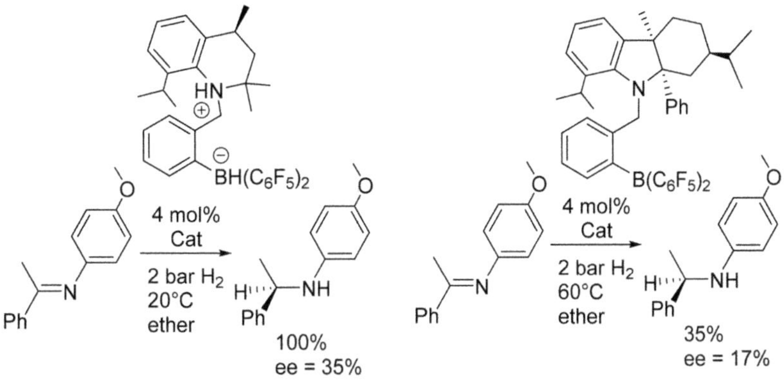

Scheme 4.12 Catalytic asymmetric reduction of imines by intramolecular chiral B/N FLPs.

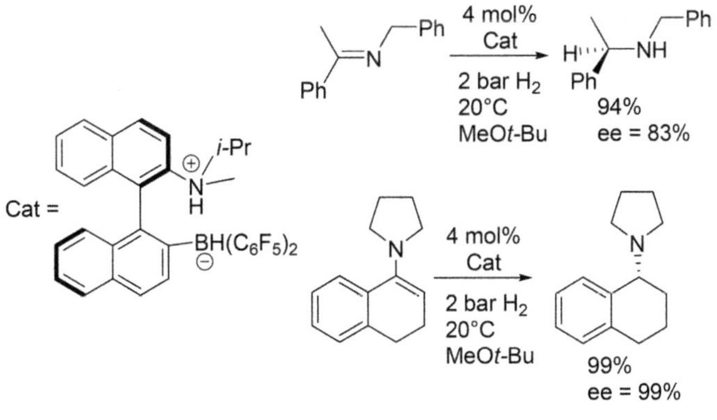

Scheme 4.13 Catalytic asymmetric reduction of imines and enamines by intramolecular chiral B/N FLPs.

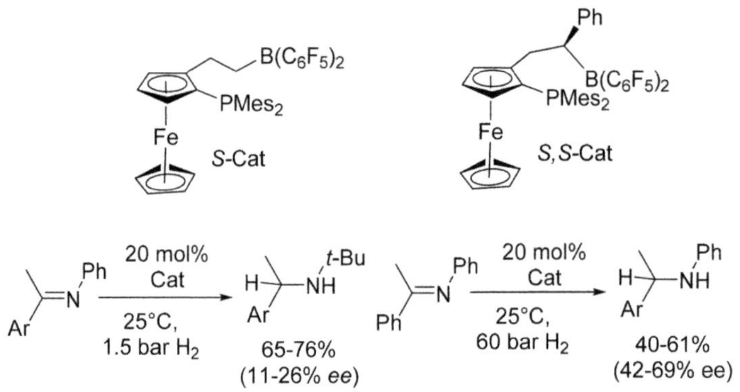

Scheme 4.14 Asymmetric reductions of imines with ferrocene derived FLPs.

Hydrogen transfer from the FLP to the enamines was found to occur in a stepwise fashion, with protonation followed by a hydride transfer. The barrier to the latter step was slightly higher than that of the protonation (23.5 and 21.8 kcal mol^{-1}, respectively), indicating that hydride delivery is rate-determining.

Erker and coworkers developed a series of chiral FLPs incorporating a ferrocene-derived linker.[22] The chiral intramolecular FLP, S-CpFeC$_5$H$_3$(PMes$_2$)(CH$_2$CH$_2$B(C$_6$F$_5$)$_2$) was shown to activate dihydrogen and mediate the reduction of selected imines and enamines (Scheme 4.14). Using 5–20 mol% catalyst, the ketimine PhC(Me)=Nt-Bu was reduced to the corresponding amine in 65–76% yield with up to 26% enantiomeric excess. In a subsequent study, Erker *et al.*[23] prepared the diastereomers of a chiral ferrocene-derived intramolecular FLP in which an additional chiral center was included in the alkyl chain bound to boron. Separation of isomers afforded the optically pure catalyst S,S-CpFeC$_5$H$_3$(PHMes$_2$)(CH$_2$CH(Ph)BH(C$_6$F$_5$)$_2$) (Scheme 4.14). This species was used in 20 mol% to mediate the hydrogenation of several ketimines employing 60 bar of dihydrogen. The chemical yields were moderate, ranging from 40–61% yields, while the product amines were obtained in enantiomeric excesses ranging from 42–69%. The corresponding reductions of N–Ph imines using the (S,R)-catalyst gave similar yields and enantiomeric excesses.[23]

4.6 Bis-borane Catalysts

A distinct and innovative alternative approach to chiral Lewis acids was developed by the Du group. In a 2013 article,[24] these authors were inspired by work on chiral olefins and envisioned the *in situ* generation of a chiral borane by hydroboration of a bis-allylbinaphthalene species with HB(C$_6$F$_5$)$_2$. Such chiral bis-boranes were thus targeted for FLP hydrogenation. In initial efforts, 5 mol% of the chiral bis-borane was used to affect the reduction of the imine PhC(Me)=NPh under 10 bar of dihydrogen at 60 °C in 15 h. The amine product was formed quantitatively, although the degree of enantioselectivity was only 20%. This was attributed to the nature of the flanking aryl groups on the binaphthyl linkage. A series of chiral bis-allylbinaphthalene precursors in which the flanking groups were altered were prepared (Scheme 4.15). At the same time, reaction conditions were optimized revealing that the use of 1.25 mol% of the diene with flanking $3,5\text{-}t\text{-}Bu_2C_6H_3$ groups in mesitylene at room temperature gave a dramatically improved enantiomeric excess of 78%. The authors went on to apply these optimized conditions to a series of FLP hydrogenations of ketimines obtaining the amine products in yields of 63–99% with enantiomeric excesses of 74–89%. The authors also confirmed that the bis-borane did not affect the epimerization of the chiral center of the amine products even under relatively forcing conditions.

A subsequent study[25] demonstrated that use of a related mono-borane-binaphthyl catalyst was also effective from the chemical hydrogenation of imines. With this catalyst, the enantiomeric excesses were not as good as the bis-borane catalysts, as they ranged from 45–89% (Scheme 4.15).

In another follow-up paper, Wei and Du extended the use of the bis-borane catalysts to the asymmetric hydrogenation of silyl-enol ethers.[26] Interestingly, generation of the bis-borane alone did not result in the reduction of the prototypical silyl-enol ether $Ph(Me_3SiO)C=CH_2$. Similarly, addition of Mes_3P to the reaction mixture was not productive. However, use of more basic phosphines in combination with the bis-borane prompted reduction. Ultimately, addition of $t\text{-}Bu_3P$ to 10 mol% of the *in situ* generated bis-borane proved to be the best conditions, giving enantiomerically pure (ee > 98%) products in >99% yield (Scheme 4.16). Again, with an established protocol this was applied to a series of silyl-enol ether substrates. Subsequent desilylation was achieved using $[Bu_4N]F$, and the resulting chiral secondary alcohols were obtained in excellent yields of 93–99% with enantiomeric excesses ranging from 88–99%.

Probing the nature of these bis-borane catalysts further, Du's group[27] examined the impact of rigidity in the chiral pocket. To this end, they developed a synthetic protocol for the alkyne analogs of the chiral binaphthyl-based diene precursors. Hydroboration of these precursors using Piers' borane afford the *in situ* generation

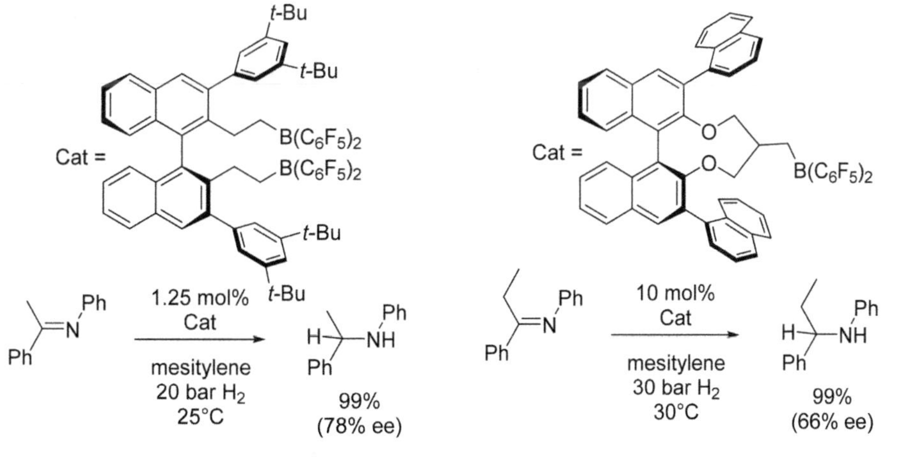

Scheme 4.15 Optimized conditions for the bis-borane and mono-borane catalyzed imine reduction.

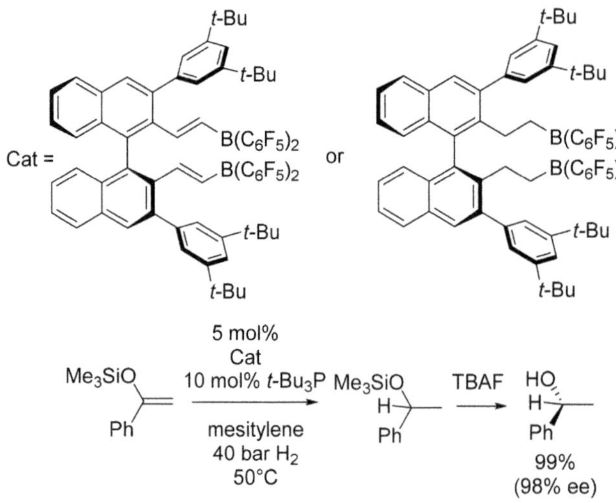

Scheme 4.16   Optimized conditions and bis-borane for silyl-enol ether reduction.

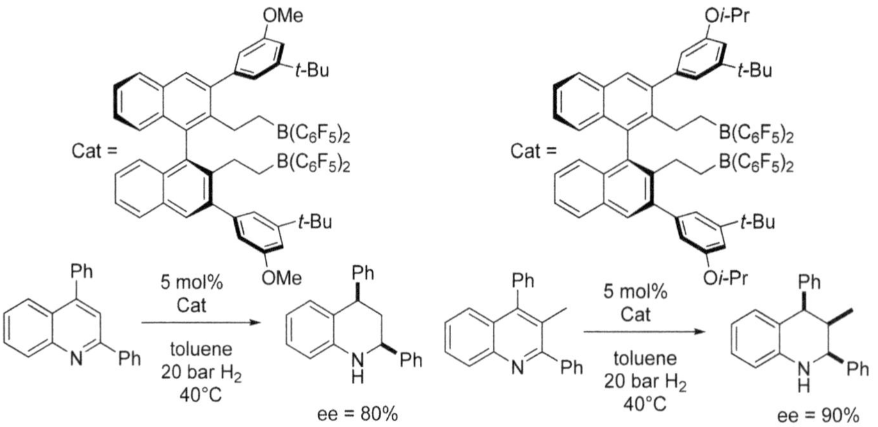

Scheme 4.17   Optimized conditions and bis-boranes for hydrogenations of N-heterocycles.

of chiral bis-alkenyl boranes (Scheme 4.17). While several variants were prepared, the optimal reduction of a silyl-enol ether again involved a protocol using t-Bu$_3$P, and the bis-borane with flanking 3,5-t-Bu$_2$C$_6$H$_3$ groups. This catalyst was used to reduce over twenty silyl enol ethers in high conversions (80–90%) and enantioselectivity ranging from 87–99%. It is noteworthy that this bis-borane derived from bis-alkyne gave lower chemical yields than those derived from the bis-alkene- binaphthalene-derived catalyst.

Continuing in the development of asymmetric reductions, the Du group next turned to substituted quinolines.[28,29] Using *in situ*-generated bis-boranes, catalyst optimization studies revealed that the optimal catalyst included flanking 3-t-Bu(5-OMe)C$_6$H$_3$ substituents. Exploiting this bis-borane, a series of 2,4- and 2,3-disubstituted quinolines were reduced in yields of 74–99% and enantiomeric excesses of 45–80% (Scheme 4.17). A similar study described the hydrogenation of 2,3,4-trisubstituted quinolines using a slightly modified catalyst, incorporating the flanking arene rings 3-t-Bu(5-Oi-Pr) C$_6$H$_3$. This afforded the *cis–cis* isomers of the products in high yields and with enantiomeric excesses ranging from 82% to 99%.[30] While efforts to extend this protocol to 2-quinolinecarboxylates gave good chemical yields, the enantioselectivities were generally very low.[31]

A marked impact of substrate structure on selectivity was subsequently reported by Du *et al.* when they applied their bis-borane catalysts to the hydrogenation 3-phenyl-1,4-benzoxazine.[32] In such cases, these catalysts effectively reduced a series of these substrate molecules in 93–99% yields. The bis-borane with flanking 3,5-t-Bu$_2$C$_6$H$_3$ groups proved most selective, affording an enantiomeric excess of only 30–42% (Scheme 4.18). Du also reported efforts to reduce a dissymmetric 1,2-diaryl-1,2-diimines using the bis-borane derived from hydroboration of the bis-allylbinaphthalene without flanking aryl substituents. While the reduction was efficient, the resulting diamine was obtained in only 10% enantiomeric excess.[33]

In a more recent 2018 work, Wang and co-workers[34] have developed an alternative synthetic route to chiral bis-boranes. Like the Du approach, the Wang synthesis involved hydroboration. However, in this case, the precursor was a bicyclic diene affording bis-boranes in which a phenyl ring and the boron substituent were on adjacent carbon atoms. Interestingly, the kinetic and thermodynamic products of hydroboration could be selectively prepared by controlling the reaction temperature. In this fashion a series of species formulated as (C$_8$H$_{10}$Ar$_2$)(B(Ar^F)$_2$)$_2$ (Ar = Ph, 4- FC$_6$H$_4$, 4-t-BuC$_6$H$_4$ 3,5(t-Bu)$_2$C$_6$H$_3$; Ar^F = C$_6$F$_5$, p-C$_6$F$_4$H) were prepared and evaluated using the imine PhC(Me)=NPh as a prototypical substrate. Indeed, all these species were effective hydrogenation catalysts when used in 2 mol% at temperatures ranging from −40 to 25 °C under 50 bar of dihydrogen. In addition, the enantiomeric excesses observed for the various catalysts ranged from 62–99%. The best catalyst (C$_8$H$_{10}$(3,5(t-Bu)$_2$C$_6$H$_3$)$_2$)

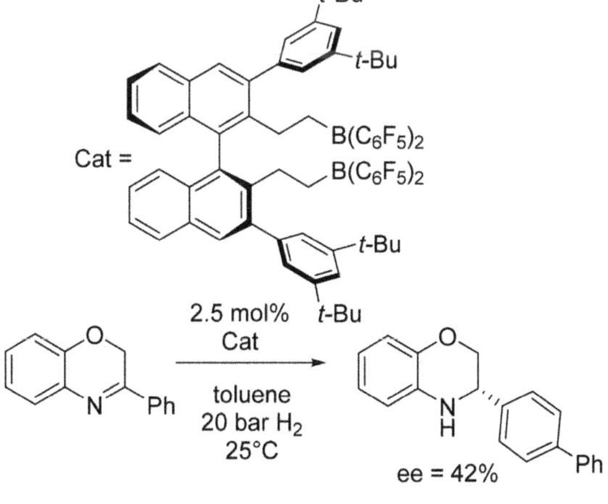

Scheme 4.18 Optimized conditions and bis-borane for phenyl-1,4-benzoxazine reductions.

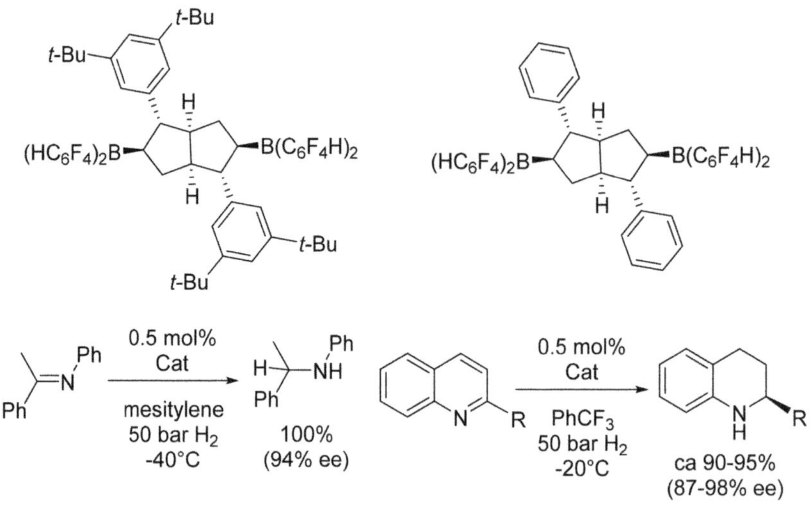

Scheme 4.19   Enantioselective reduction of an imine by a bis-borane catalyst.

$(B(p\text{-}C_6F_4H)_2)_2$ was even effective at 0.5 mol% and was subsequently employed to affect the asymmetric reduction of a broad spectrum of aryl imines at $-40\,^{\circ}$C (Scheme 4.19). In over 30 examples, the majority of the enantioselectivities seen were greater than 90%. In contrast, prochiral dialkyl imines were not reduced under these conditions and the prochiral diphenyl imine was reduced in only 44% yield with an enantiomeric excess of only 3%.

In a subsequent study, Wang and coworkers[35] employed 2–5 mol% of the variant of the above bis-borane catalyst, $C_8H_{10}(Ph_2)(B(p\text{-}C_6F_4H)_2)_2$, to affect the hydrogenation of a series of 45 alkyl and aryl substituted quinolines (Scheme 4.19). The N-containing rings of the substrates were effectively reduced in high yields with enantioselectivities ranging from 87–99%.

4.7   Chiral Intermolecular FLPs

In a 2020 advance, the Du group[36] recognized the key role hydrogen bonding plays in the mechanism of carbonyl reduction (see Section 3.12). These authors cleverly adapted this aspect for asymmetric induction employing a borane in the presence of a chiral oxazoline Lewis base. Optimization of the oxazoline substituents and reaction conditions led to the use of 10 mol% of the borane $B(p\text{-}C_6F_4H)_3$ with 20 mol% of the oxazoline $(PhCH)_2NOCt\text{-}Bu$ at 30 $^{\circ}$C under 40 bar of dihydrogen pressure. These conditions allowed the reduction of 30 prochiral ketones in 43–97% yielding enantioselectivities ranging from 50–86% (Scheme 4.20). Similarly, this strategy was applied to the reduction of 34 1-tetralone-derived enones using the FLP $(PhCH)CH_2NOCEt_3/B(p\text{-}C_6F_4H)_3$ affording yields of 86–99% with enantiomeric excesses of 63–92% (Scheme 4.20).

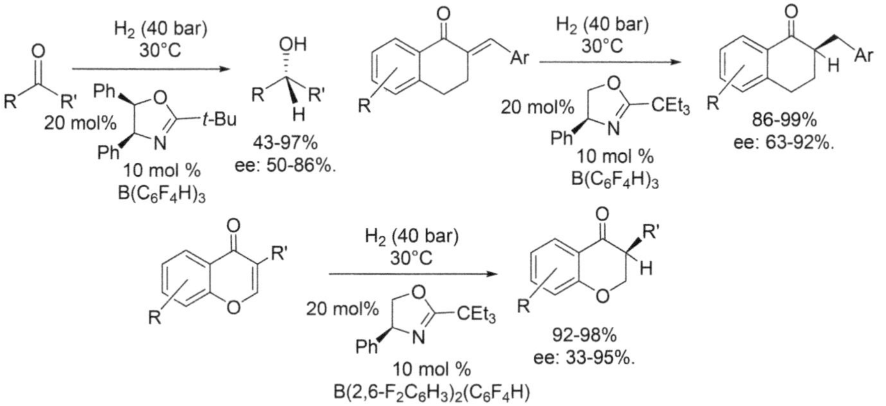

Scheme 4.20 Asymmetric reduction using chiral oxazoline and an achiral borane.

In addition, a series of 16 3-substituted chromones were also reduced with yields ranging from 92–98% and enantioselectivities of 33–95% (Scheme 4.20). In the latter case, reductions were optimized using the borane $B(2,6-F_2C_6H_3)_2(C_6F_4H)$.

4.8 Implications

The above results illustrate the remarkably rapid evolution of asymmetric FLPs catalysts over about a decade. While early efforts offered meager enantioselectivities, several classes of highly efficient catalysts have emerged. While each required structural optimization, chiral boranes, chiral intramolecular FLPs, and chiral bis-boranes have all proved highly effective and selective for asymmetric hydrogenations. A more recent development has been the use of chiral FLPs derived from chiral oxazoline and borane. As a result of all these developments, a comparatively broad range of substrates has been reduced with high selectivities.

While these results establish the viability of asymmetric FLP hydrogenations, there is no doubt that this aspect of FLP chemistry will continue to evolve. New strategies to chiral catalysts that exploit readily available precursors would be desirable. In addition, catalysts that are easily handled in air, exhibit broad functional group tolerance, and of course provide high enantioselectivity remain important targets.

The advent of metal-free asymmetric reductions and the remarkable success to date would also appear to provide a highly desirable protocol for the synthesis of small molecule drugs or pharmaceuticals. Certainly, such technologies offer strategies to selected chiral centers without the simultaneous introduction of transition metal residue which is both toxic and costly to remove. Whether or not chiral FLP catalysts find commercial applications or not, time will tell. At the very least, these advances offer new reduction protocols that are worthy of assessment as production strategies for new materials. As such, chiral FLP catalysts provide a new tool in the synthetic arsenal of the organic chemist.

References

1. D. T. Hog and M. Oestreich, *Eur. J. Org. Chem.*, 2009, 5047–5056.
2. A. G. Massey and A. J. Park, *J. Organomet. Chem.*, 1964, **2**, 245–250.
3. N. Millot, C. C. Santini, B. Fenet and J. M. Basset, *Eur. J. Inorg. Chem.*, 2002, 3328–3335.
4. F. Focante, I. Camurati, D. Nanni, R. Leardini and L. Resconi, *Organometallics*, 2004, **23**, 5135–5141.
5. V. Sumerin, F. Schulz, M. Nieger, M. Leskela, T. Repo and B. Rieger, *Angew. Chem., Int. Ed.*, 2008, **47**, 6001–6003.
6. J. M. Farrell, Z. M. Heiden and D. W. Stephan, *Organometallics*, 2011, **30**, 4497–4500.
7. G. Eros, H. Mehdi, I. Papai, T. A. Rokob, P. Kiraly, G. Tarkanyi and T. Soós, *Angew. Chem., Int. Ed.*, 2010, **49**, 6559–6563.
8. Z. M. Heiden and D. W. Stephan, *Chem. Commun.*, 2011, **47**, 5729–5731.
9. K. Unverhau, G. Lubbe, B. Wibbeling, R. Fröhlich, G. Kehr and G. Erker, *Organometallics*, 2010, **29**, 5320–5329.
10. D. W. Stephan, S. Greenberg, T. W. Graham, P. Chase, J. J. Hastie, S. J. Geier, J. M. Farrell, C. C. Brown, Z. M. Heiden, G. C. Welch and M. Ullrich, *Inorg. Chem.*, 2011, **50**, 12338–12348.
11. N. A. Sitte, L. Koering, P. W. Roesky and J. Paradies, *Org. Biomol. Chem.*, 2020, **18**, 7321–7325.
12. D. J. Chen and J. Klankermayer, *Chem. Commun.*, 2008, 2130–2131.
13. D. J. Parks, W. E. Piers and G. P. A. Yap, *Organometallics*, 1998, **17**, 5492–5503.
14. D. Chen, Y. Wang and J. Klankermayer, *Angew. Chem., Int. Ed.*, 2010, **49**, 9475–9478.
15. D. Chen, V. Leich, F. Pan and J. Klankermayer, *Chem. - Eur. J.*, 2012, **18**, 5184–5187.
16. G. Ghattas, D. Chen, F. Pan and J. Klankermayer, *Dalton Trans.*, 2012, **41**, 9026–9028.
17. A. Hamza, K. Sorochkina, B. Kotai, K. Chernichenko, D. Berta, M. Bolte, M. Nieger, T. Repo and I. Papai, *ACS Catal.*, 2020, **10**, 14290–14301.
18. S. Schwendemann, S. Oishi, S. Saito, R. Fröhlich, G. Kehr and G. Erker, *Chem. - Asian J.*, 2013, **8**, 212–217.
19. M. Lindqvist, K. Axenov, M. Nieger, M. Raisanen, M. Leskela and T. Repo, *Chem. - Eur. J.*, 2013, **19**, 10412–10418.
20. V. Sumerin, K. Chernichenko, M. Nieger, M. Leskelä, B. Rieger and T. Repo, *Adv. Synth. Catal.*, 2011, **353**, 2093–2110.
21. M. Lindqvist, K. Borre, K. Axenov, B. Kótai, M. Nieger, M. Leskela, I. Pápai and T. Repo, *J. Am. Chem. Soc.*, 2015, **137**, 4038–4041.
22. X. Wang, G. Kehr, C. G. Daniliuc and G. Erker, *J. Am. Chem. Soc.*, 2014, **136**, 3293–3303.
23. K. Y. Ye, X. W. Wang, C. G. Daniliuc, G. Kehr and G. Erker, *Eur. J. Inorg. Chem.*, 2017, 368–371.
24. Y. B. Liu and H. F. Du, *J. Am. Chem. Soc.*, 2013, **135**, 6810–6813.
25. X. Liu, T. Liu, W. Meng and H. Du, *Org. Biomol. Chem.*, 2018, **16**, 8686–8689.
26. S. M. Wei, X. Q. Feng and H. F. Du, *J. Am. Chem. Soc.*, 2014, **136**, 12261–12264.
27. X. Y. Ren, G. Li, S. M. Wei and H. F. Du, *Org. Lett.*, 2015, **17**, 990–993.
28. Z. H. Zhang and H. F. Du, *Angew. Chem., Int. Ed.*, 2015, **54**, 623–626.
29. Z. H. Zhang and H. Du, *Org. Lett.*, 2015, **17**, 6266–6269.
30. Z. H. Zhang and H. F. Du, *Org. Lett.*, 2015, **17**, 2816–2819.
31. C. Han, E. Zhang, X. Feng, S. Wang and H. Du, *Tetrahedron Lett.*, 2018, **59**, 1400–1403.
32. S. Wei, X. Feng and H. Du, *Org. Biomol. Chem.*, 2016, **14**, 8026–8029.
33. X. Zhu and H. Du, *Org. Lett.*, 2015, **17**, 3106–3109.
34. X. S. Tu, N. N. Zeng, R. Y. Li, Y. Q. Zhao, D. Z. Xie, Q. Peng and X. C. Wang, *Angew. Chem., Int. Ed.*, 2018, **57**, 15096–15100.
35. X. Li, J. J. Tian, N. Liu, X. S. Tu, N. N. Zeng and X. C. Wang, *Angew. Chem., Int. Ed.*, 2019, **58**, 4664–4668.
36. B. Gao, X. Feng, W. Meng and H. Du, *Angew. Chem., Int. Ed.*, 2020, **59**, 4498–4504.

5 Structural Variations of FLPs

5.1 Chapter Overview

Since boranes are perceived as the quintessential Lewis acid, it is not surprising that they have been central to the discovery and implementation of FLP chemistry. Nonetheless, these findings stimulate several questions. One issue at the forefront is the generality of the concept. For example, can FLP chemistry be generalized to other Lewis acid/base combinations? Can other types of acidic molecules be used in dihydrogen activation and catalytic hydrogenation? If so, what are the changes that are required to allow these systems to be useful catalytically? Indeed, what are the general features required to achieve the 'frustration' required for FLP reactivity? In this chapter, we address some of these questions by surveying a broad range of systems that have been shown to react as FLPs with dihydrogen. Indeed, the broad range of systems described herein demonstrates that this concept extends across the main group and more generally across the periodic table. These findings confirm that the notion of frustration and FLP reactivity is not a niche area but rather a broader chemical concept, that proves useful for the design of novel catalysts for hydrogenation.

5.2 Boron Cations

One of the first strategies to an alternate class of catalyst involved the use of a boron cation. This idea seemed attractive, as the cationic charge would provide Lewis acidity while one could envision generating such a cation from a Lewis base adduct of a secondary borane. In addition, the Lewis base–borane adduct would provide an air-stable precursor to the cation. To this end, the N-heterocyclic carbene (NHC)–borane adduct, $(C_3H_2(Ni\text{-}Pr)_2)(HB(C_8H_{14}))$ was prepared. This species is indeed air-stable and robust, while subsequent treatment of the adduct $(C_3H_2(Ni\text{-}Pr)_2)(HB(C_8H_{14}))$ with $[Ph_3C]$ $[B(C_6F_5)_4]$ resulted in the abstraction of the boron-bound hydride, generating Ph_3CH and the borenium cation salt $[(C_3H_2(Ni\text{-}Pr)_2)B(C_8H_{14})][B(C_6F_5)_4]$.[1] The ability of this

A Primer in Frustrated Lewis Pair Hydrogenation: Concepts to Applications
By Douglas W. Stephan
© Douglas W. Stephan 2022
Published by the Royal Society of Chemistry, www.rsc.org

species to act as Lewis acid in FLP chemistry was affirmed in a reaction with *t*-Bu$_3$P and dihydrogen, as this resulted in the regeneration of the NHC–borane adduct and [*t*-Bu$_3$PH][B(C$_6$F$_5$)$_4$] (Scheme 5.1).

Furthermore, 1–5 mol% of the borenium cation was employed to mediate the hydrogenation of several imines, enamines as well as methylquinoline at room temperature under 102 bar of dihydrogen.[1] Such reactions were generally quantitative within 1 to 4 hours. Efforts to modify the borane used in such cations were undertaken. For example, attempts to include a more electrophilic boron center prompted the synthesis of (C$_3$H$_2$(N*t*-Bu)$_2$)BH(C$_6$F$_5$)$_2$.[2] However, the corresponding borenium cation was not isolable although the reaction of this NHC–borane adduct with HNTf$_2$ inferred the transient generation of the corresponding cation. However, the zwitterionic product, C$_3$H$_2$(N*t*-Bu)(NCMe$_2$CH$_2$)B(C$_6$F$_5$)$_2$, was isolated, resulting from the C–H activation of one of the *t*-butyl substituents (Scheme 5.2). This observation suggests that the heightened Lewis acidity of this transient borenium center precludes utility in catalysis, as it is deactivated by intramolecular C–H bond activation.

Nonetheless, to further examine the impact of structural variations on the catalytic efficiency of borenium cations, a series of cations of the form [(NHC)B(C$_8$H$_{14}$)][B(C$_6$F$_5$)$_4$] were prepared using readily accessible carbene donors.[2] These cations were used in a prototypical imine reduction of PhCH=N*t*-Bu at room temperature and 102 bar of hydrogen. Quenching of the reactions after 30 minutes revealed a counter-intuitive observation. Carbenes with bulky substituents on nitrogen such as *i*-Pr, *t*-Bu, Mes, 2,6-*i*-Pr$_2$C$_6$H$_3$ exhibited relatively poor reactivity, whereas less sterically demanding substituents offered improved reactivity. Ultimately, the species [(Cl$_2$C$_3$(NMe)$_2$)B(C$_8$H$_{14}$)][B(C$_6$F$_5$)$_4$] proved to be the most effective, providing quantitative reduction of the imine

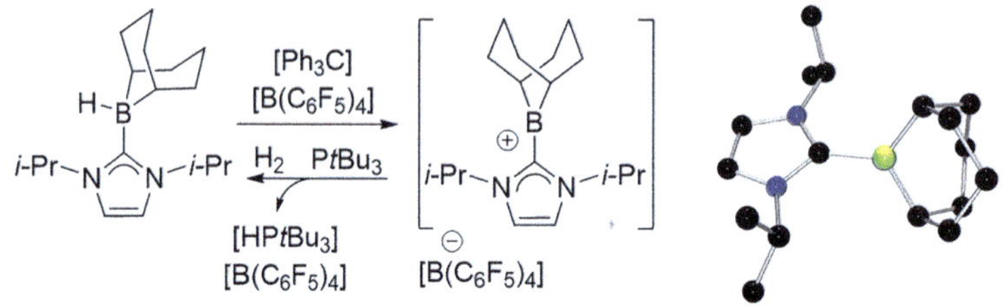

Scheme 5.1   Reactivity of a borenium cation; POV-ray depiction of the cation, [(C$_3$H$_2$(N*i*-Pr)$_2$)B(C$_8$H$_{14}$)]$^+$.

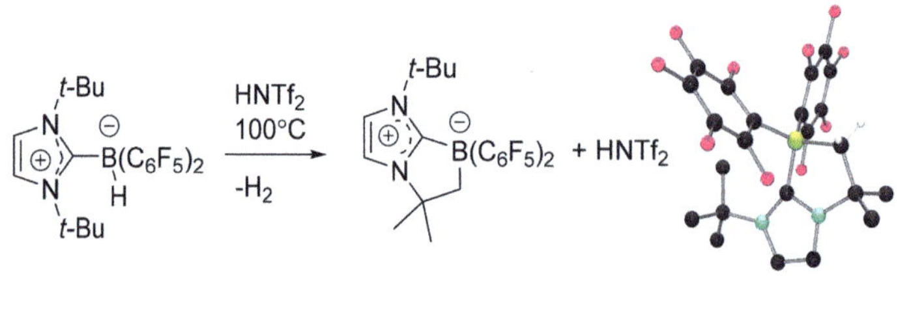

Scheme 5.2   Efforts to target a borenium cation incorporating C$_6$F$_5$ groups on boron.

in 30 minutes with as little as 0.15 mol% catalyst loading. This reactivity corresponded to a turn-over frequency (TOF) of 940 h^{-1} which, while among the most active of FLP catalysts known, is significantly lower than those achieved by transition metal-based catalysts. This borenium catalyst was also used to reduce a series of imines and enamines, exhibiting tolerance for the inclusion of halogens, esters, and ethers in the substrate molecules.[2]

Mechanistically this system is remarkably straightforward. The borenium cation and the imine substrate act as an FLP generating the NHC–borane adduct and the transient iminium cation, which is of sufficient Lewis acidity to abstract the hydride from boron. This provides the amine product and regenerates the borenium cation which is available for further activation of dihydrogen (Scheme 5.3).[1,2] Further experimental studies were consistent with the activation of dihydrogen being the rate-determining step.

The Crudden group[3] expanded the scope of carbene-stabilized borenium cations to include those derived from several triazole-derived mesoionic N-heterocyclic carbenes (MIC). In a similar fashion to those described above, compounds of the form [(MIC)HB(C$_8$H$_{14}$)] were prepared, and abstraction of the hydride with [Ph$_3$C][B(C$_6$F$_5$)$_4$] or B(C$_6$F$_5$)$_3$ generated the corresponding salts (Scheme 5.4). These cations were employed in the catalytic hydrogenation of the imines PhCH=Nt-Bu and PhC(Me)=NPh under varying dihydrogen pressures. The efficacies of these catalysts were compared with the NHC-derived analogs as well as a borenium cation incorporating an abnormally bound NHC. In general, the MIC derivatives proved more active, reducing the imines even at 1 bar of dihydrogen and room temperature (Scheme 5.4). The authors also went on to use 10 mol% of the species [(C$_2$N$_3$Ph$_2$)B(C$_8$H$_{14}$)][B(C$_6$F$_5$)$_4$] to affect the reductions of a series of quinoline and pyridine derivatives. At 102 bar of dihydrogen at room temperature, the yields of these reductions ranged from 28–98% depending on the nature of the substitution on the N-heterocycle. Interestingly, the reduction of 2,6-disubstituted pyridines resulted in the formation of single diastereomers.

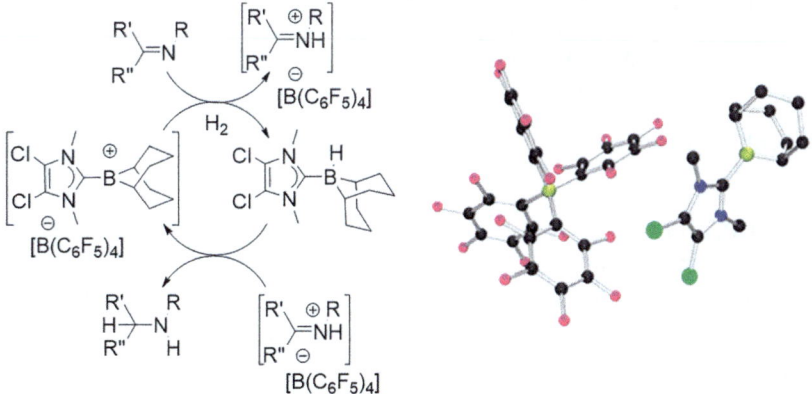

Scheme 5.3 Mechanism of borenium catalysis for imine reduction and POV-ray depiction of the optimized borenium cation salt.

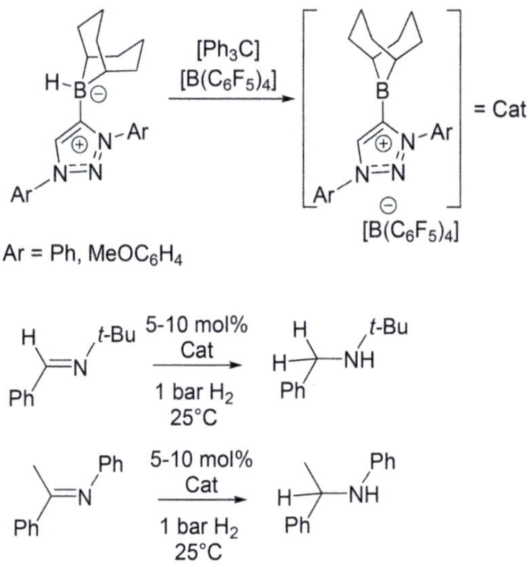

Scheme 5.4 Examples of imine reductions by borenium cations derived from MIC ligands.

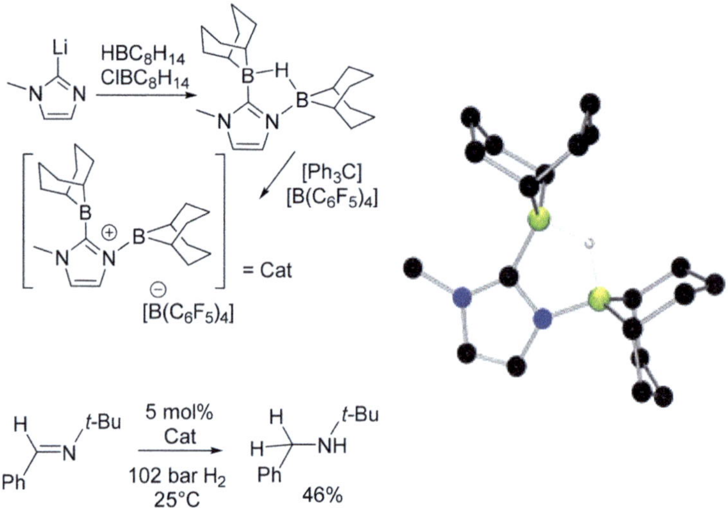

Scheme 5.5 Synthesis and reactivity of a unique bis-borane cation.

A species containing two boron centers was prepared from the reaction of the lithium salt of methyl imidazole with $HB(C_8H_{14})$ and $ClB(C_8H_{14})$,[4] resulting in a hydride-bridged di-borane derivative. Abstraction of the hydride generated a unique cation containing two three-coordinate boron centers, $[(C_3H_2N_2Me)(B(C_8H_{14}))_2][B(C_6F_5)_4]$ (Scheme 5.5). Using 5 mol% of this species as a catalyst, the hydrogenation of PhCH=Nt-Bu was performed using 102 bar of dihydrogen pressure at room temperature. After 24 h this species proved to provide only moderate activity, as the amine product was isolated in 46% yield.

Ingleson and coworkers[5] also probed the FLP reactivity of borenium cations. These authors prepared [C$_6$H$_4$O$_2$BPt-Bu$_3$][AlCl$_4$] and showed that, in the presence of an additional equivalent of phosphine, this mixture affected the cleavage of dihydrogen yielding [HPt-Bu$_3$][AlCl$_4$] and C$_6$H$_4$O$_2$BH(Pt-Bu$_3$) (Scheme 5.6). The latter species is a weak adduct that established an equilibrium with free borane and phosphine. These authors also studied related borenium cations of the form [(C$_5$H$_3$Me$_2$N)BClR][B(C$_6$H$_3$Cl$_2$)$_4$] (R = Cl, Ph), derived from lutidine and BCl$_3$. In combination with PMes$_3$, these cations affected the activation of D$_2$ generating the phosphonium salt [DPMes$_3$][B(C$_6$H$_3$Cl$_2$)$_4$] and the lutidine–borane adduct (C$_5$H$_3$Me$_2$N)BDClPh (Scheme 5.6). At the same time, the additional by-products (C$_5$H$_3$Me(CH$_2$)N)BClPh and [HPMes$_3$][B(C$_6$H$_3$Cl$_2$)$_4$] are formed from the activation of a methyl group of lutidine.

The Bourissou group[6] probed borenium cations in which a phosphine donor is linked *via* a naphthalene fragment to boron and are thus formulated as [Ph$_2$P(C$_{10}$H$_6$)BMes] X (X = GaCl$_4$, NTf$_2$). These cations react with dihydrogen in the presence of *t*-Bu$_3$P to form Ph$_2$P(C$_{10}$H$_6$)BHMes and [*t*-Bu$_3$PH]X whereas in the absence of additional phosphine, loss of a mesityl group *via* protonation afforded the borane Ph$_2$P(C$_{10}$H$_6$)BHNTf$_2$ (Scheme 5.7). This latter reaction requires heating to 80 °C, consistent with computations that showed an intermediate dihydrogen complex, reacting *via* a transition state that permits protonation of the mesityl carbon.

In 2016, a collaboration among the Stephan, Melen, and Crudden groups[7] undertook efforts to adapt borenium catalysts for asymmetric reductions. To this end, chirality was incorporated *via* several strategies. N-Heterocyclic carbenes or mesoionic carbenes that have chiral backbones or chiral substituents on the nitrogen atoms were employed.

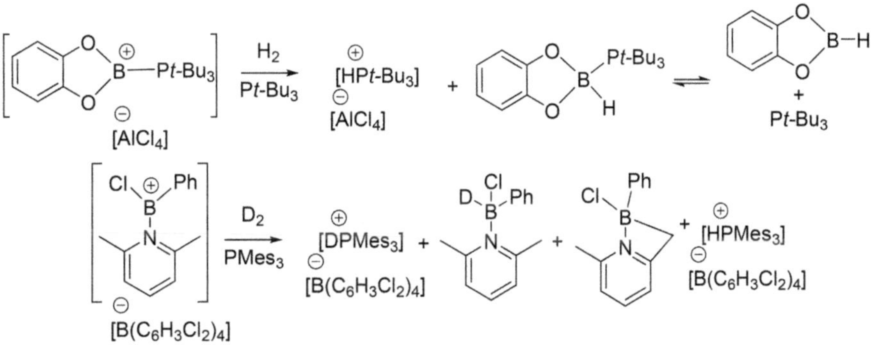

Scheme 5.6 Ingleson's studies of borenium cations.

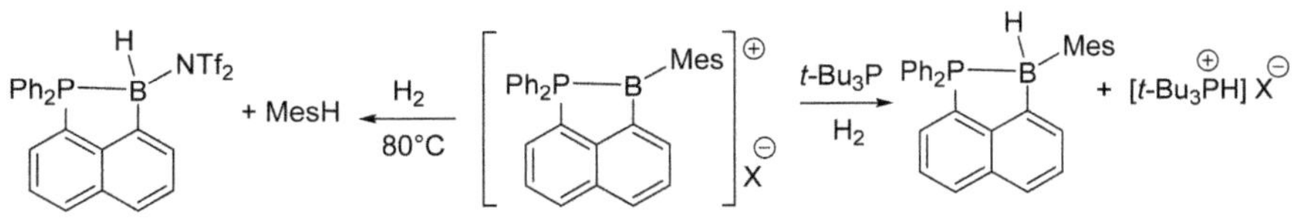

Scheme 5.7 Reactivity of [Ph$_2$P(C$_{10}$H$_6$)BMes]X with dihydrogen.

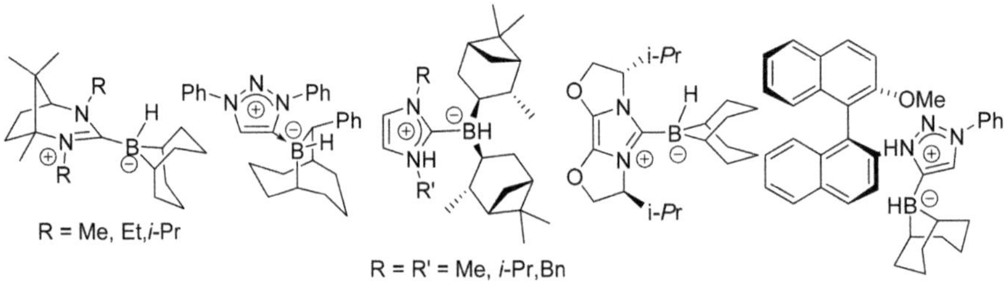

R = Me, Et,*i*-Pr

R = R' = Me, *i*-Pr,Bn

Figure 5.1 Examples of chiral precursors to borenium cations.

Alternatively, several examples in which chiral substituents were incorporated on boron were also prepared. While the corresponding borenium cations could be prepared, generally use of 5 mol% of these species showed poor reactivity as a catalyst for the hydrogenation of the prochiral imine PhC(Me)=NPh, even at 102 bar of dihydrogen pressure. Only the cations incorporating the chiral py-box carbene, [(C$_2$(OC$_2$H$_3$ *i*-Pr)$_2$N$_2$C)B(C$_8$H$_{14}$)]$^+$, or the mesoionic carbene derivative with a binaphthyl substituent on nitrogen, [(C$_{20}$H$_{12}$(OMe)N$_2$(NPh)CHC)B(C$_8$H$_{14}$)]$^+$ (Figure 5.1) were able to affect the quantitative reduction of this prototypical imine. However, in both cases, the degree of enantioselectivity was poor rising only to 6 or 7% enantiomeric excess. While other borenium cations exhibited higher selectivity, chemical yields were poor. These observations suggest that there is a significant challenge in achieving a borenium cation that is sufficiently bulky to induce high selectivity and yet accessible enough to provide an efficient catalyst.

5.3   Nucleophilic Boron

Another approach to unique FLPs was considered based on the 2006 synthesis of boryl anions.[8] While that work continues to garner significant interest as a fundamental finding representing an unprecedented umpolung in boron reactivity,[9] the use of the boryl anion as a nucleophile in FLP chemistry was not probed until 2015 when Wang and coworkers[10] used an FLP derived from the combination of Li[pinBB(Ph)pin] and B(C$_6$F$_5$)$_3$. This combination was shown to activate dihydrogen at ambient temperature affording PhBpin, Li[HB(C$_6$F$_5$)$_3$] and B$_2$pin$_2$. However, computations infer that the nucleophilic sp^2 boryl fragment of Li[pinBB(Ph)pin] is not directly involved in the cleavage of the H–H bond.

In a related sense, Zhao and coworkers[11] probed the chemistry of the 1,3,2,5-diazadiborinine bis-boron heterocycle (PhB)$_2$(NCH$_2$CMe$_2$OC)$_2$ prepared by Kinjo *et al.*[12] Recognizing the polar nature of this species, it can be viewed as containing an electron-deficient B(III) center and an electron-rich B(I) center. From that perspective, this species is an intramolecular FLP. While these authors showed that this heterocycle indeed reacts as an FLP with a series of small organic molecules, the reaction with dihydrogen to affect heterolytic cleavage generating (PhBH)$_2$(NCH$_2$CMe$_2$OC)$_2$ was limited to computations (Scheme 5.8), and no experimental results were reported.

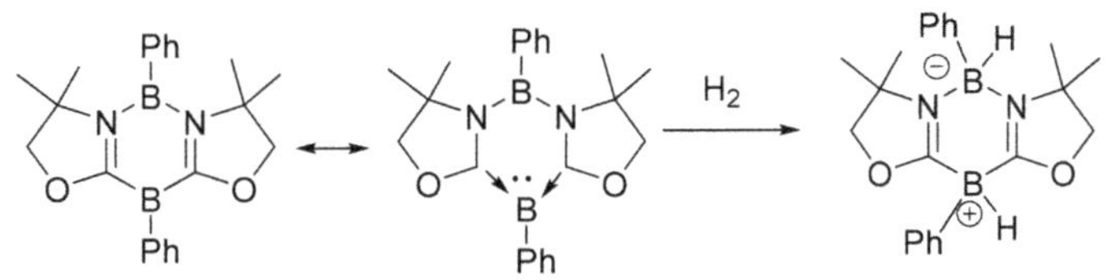

Scheme 5.8 Computed reaction of B/B FLP with dihydrogen.

5.4 Alanes, Galanes and Indanes

Another avenue for Lewis acid variation is the rather obvious extension to heavier group (III) compounds. While early works describing the reactions of FLPs with alkynes did include the use of the aluminium Lewis acid $Al(C_6F_5)_3$,[13] the first study examining reactions of this Lewis acid with dihydrogen only appeared in 2012.[14] In that case, combinations of $Al(C_6F_5)_3$ with two equivalents of PR_3 (R = *t*-Bu, Mes) affected the heterolytic cleavage of dihydrogen, providing the salts $[R_3PH][(\mu\text{-H})(Al(C_6F_5)_3)_2]$ (Scheme 5.9). It is thought that this reaction proceeds by analogy to the boron system; however, interaction of the transient anion $[HAl(C_6F_5)_3]$ with the Lewis acidic $Al(C_6F_5)_3$ affords the observed anion with the hydride bridging the two aluminium centers.

Probing the possibility of olefin reduction, the salts $[R_3PH][(\mu\text{-H})(Al(C_6F_5)_3)_2]$ were reacted with ethylene or cyclohexene.[15] This results in the formation of the products $[R_3PH][Al(C_6F_5)_4]$ and $R'Al(C_6F_5)_2$ (Scheme 5.9) which are formed by a redistribution reaction involving the initial intermediate $[HPR_3][R'Al(C_6F_5)_3]$ with $Al(C_6F_5)_3$. Interestingly, the mono-aluminium hydride salt $[Et_4N][HAl(C_6F_5)_3]$ alone did not react with olefins. However, in the presence of additional $Al(C_6F_5)_3$, the reaction proceeded, inferring that alane activation of the olefin is required.[15] This notion was further supported by crystallographic characterization of the complex, $(C_6H_{10})Al(C_6F_5)_3$ in which the olefin interacts with aluminium in an η^2-fashion. Efforts to transform the above stoichiometric chemistry to catalysis seemed promising as salts of the form $[R_3PH][Al(C_6F_5)_4]$ can protonate the alkyl-aluminates, $[Et_4N][R'Al(C_6F_5)_3]$, liberating alkanes. However, the P/Al FLPs in the presence of dihydrogen and ethylene generated the anion $[R_3PH]$ $[EtAl(C_6F_5)_3]$, which in the presence of free $Al(C_6F_5)_3$ underwent a redistribution reaction giving $EtAl(C_6F_5)_2$ and the anion $[Al(C_6F_5)_4]$ rather than the liberation of ethane *via* protonation of the anion by the phosphonium cation.

Extension of FLP chemistry to gallium and indium was also reported several years later.[16] Indeed reaction of *t*-Bu$_3$P with two equivalents of $Ga(C_6F_5)_3$ under dihydrogen (1 bar) afforded the corresponding salts $[t\text{-Bu}_3PH][(Ga(C_6F_5)_3)_2(\mu\text{-H})]$ (Scheme 5.10). Similarly, the analogous reaction of $In(C_6F_5)_3$ with Mes$_3$P under dihydrogen afforded $[Mes_3PH]$ $[(In(C_6F_5)_3)_2(\mu\text{-H})]$, demonstrating that Ga and In behave similarly to Al. However, $Ga(C_6F_5)_3$ and $In(C_6F_5)_3$ were employed in the catalytic hydrogenation of PhCH=N*t*-Bu. Using 5 mol% of these Lewis acids, the imine was reduced under 4 bar of dihydrogen at 130 °C in 12 h. Under these conditions, the gallium catalyst afforded quantitative reduction, while the indium analog achieved 85% conversion (Scheme 5.10).

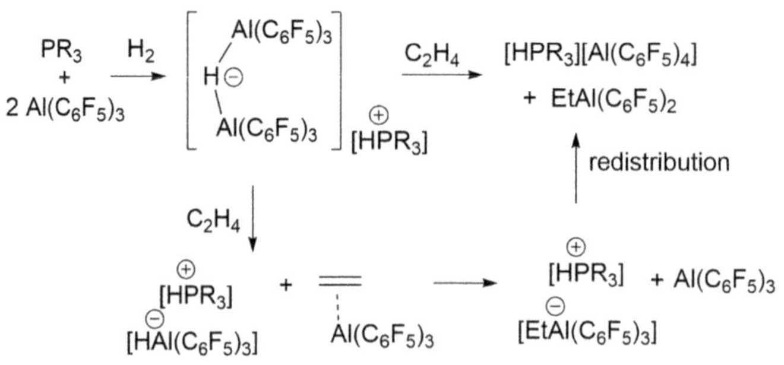

Scheme 5.9   Activation of dihydrogen by Al(C$_6$F$_5$)$_3$/PR$_3$ and reaction with ethylene.

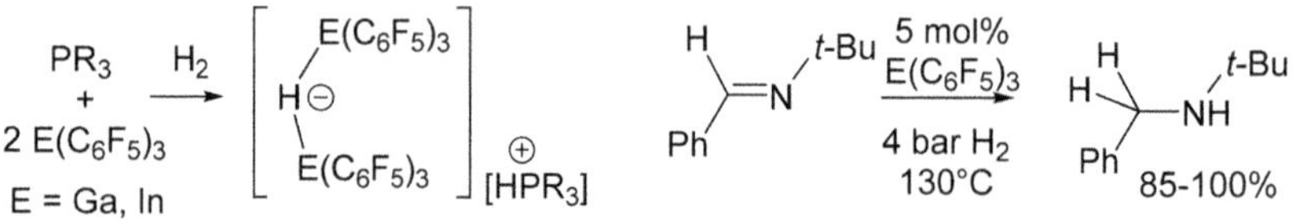

Scheme 5.10   Stoichiometric and catalytic reactions of E(C$_6$F$_5$)$_3$ (E = Ga, In).

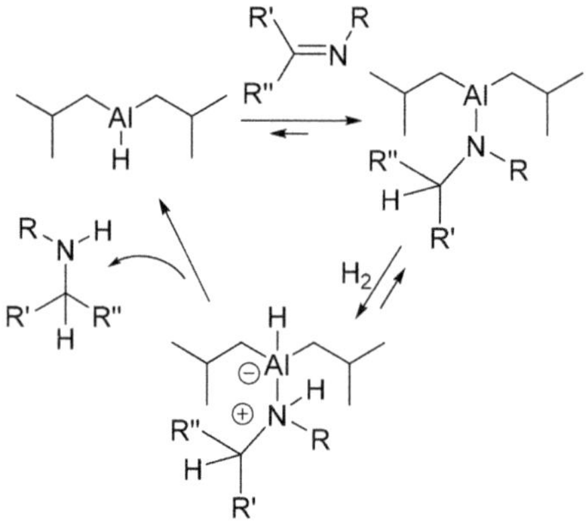

Scheme 5.11   Reduction of imines *via* a hydroalumination/hydrogenolysis mechanism.

Efforts to apply other alanes to catalysis were motivated by the pragmatic notion that simple alkyl aluminium compounds such as di-isobutyl aluminium hydride (DIBAL) and tri-isobutyl aluminium hydride (TIBAL) were available on a large scale. To probe the potential of these species in catalysis, such compounds were tested in the hydrogenation of a variety of imines. At 100 °C and 100 bar of dihydrogen, these Lewis acids were found to be effective in providing amine products in 48 h with yields ranging from 16–100% depending on the substituents on the imine.[17] In contrast to the anticipated FLP mechanism, experimental evidence inferred that DIBAL reacts with the imine *via* hydroalumination affording a transient amido-alane. This species undergoes hydrogenolysis to regenerate DIBAL and provide the amine product (Scheme 5.11). This view was

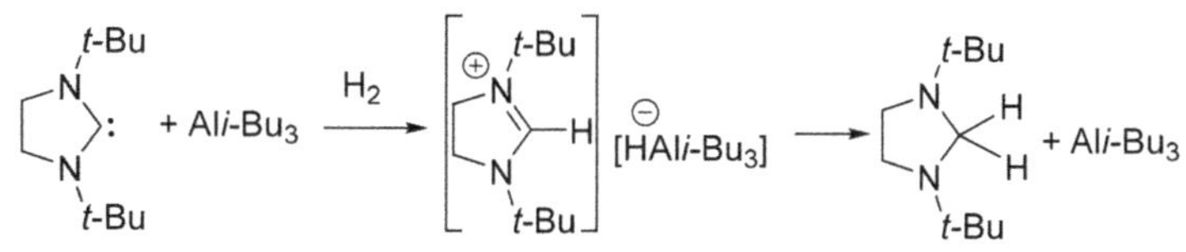

Scheme 5.12 Reaction of carbene and alane with dihydrogen.

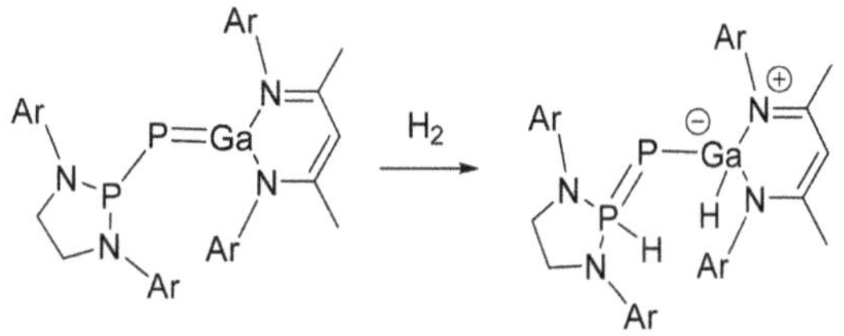

Scheme 5.13 Reaction of an intramolecular Ga/P FLP with dihydrogen.

supported by the isolation of the dimeric species $[(Me_2CHCH_2)_2Al(NPh(CH_2C_6H_4Br))]_2$ at room temperature, as higher temperatures are required for dissociation of the dimer and hydrogenolysis. Using TIBAL as the precursor, the reaction is thought to follow a similar path, as hydrogenolysis of TIBAL at elevated temperatures was shown to generate the active DIBAL catalyst with loss of isobutylene.

Dihydrogen activation has also been observed for the FLP system derived from the N-heterocyclic carbene $C_3H_4(Nt\text{-Bu})_2$ and $Al(i\text{-Bu})_3$. In this case, the transient salt is not isolated. Rather, the anion delivered hydride to the cation yielding $C_3H_6(Nt\text{-Bu})_2$ (Scheme 5.12).[18] The analogous behavior was also observed for $GaMe_3$ and $InMe_3$.[19] A very recent report has also described the scrambling of HD gas to dihydrogen and D_2, effected by a combination of Ph_2NMe and $Al(i\text{-Bu})_3$ as well as the intramolecular Al/N FLP $C_6H_4(NC_5H_6Me_4)(AlH_2)$.[20]

Goicoechea and coworkers[21] have recently reported the synthesis of rare examples of a species containing a gallium–phosphorus double bond. These species formulated as $HC(C(Me)NR)_2Ga=PP((RN)_2C_2H_4)$ and $HC(C(Me)NR)_2Ga=PP((RN)_2C_2H_2)$ (R = 2,6-i-$Pr_2C_6H_3$) were prepared by reaction of $OCPP((RN)_2C_2H_x)$ (x = 2,4) with the gallium(I) precursor. These species act as intramolecular FLPs, reacting with dihydrogen to protonate phosphorus and deliver hydride to gallium (Scheme 5.13).

5.5 Aluminium Cations

Aluminium cations have only very recently been examined in FLP hydrogenation catalysis. Harder and coworkers[22] prepared a series of aluminium cations of the general form $[(HC(C(R)N(2,6\text{-}R'_2C_6H_3)_2)AlX][B(C_6F_5)_4]$ (R = Me, t-Bu, R' = i-Pr, i-Pe, X = H, Me) and examined the ability of these species to mediate the hydrogenation of imines. Interestingly while $[(HC(C(Me)N(2,6\text{-}i\text{-}Pr_2C_6H_3)_2)AlMe][B(C_6F_5)_4]$ formed an unreactive adduct with the imine $PhC(H)=N\text{-}t\text{-Bu}$ (Scheme 5.15), the analogous species with t-Bu substituents

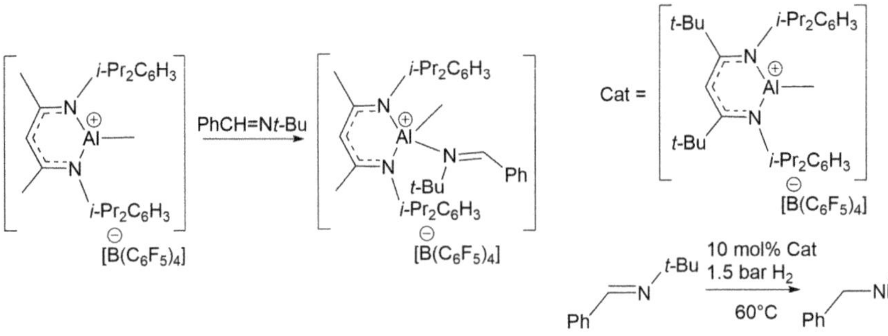

Scheme 5.14 Reactions of Al-cations with imines.

on the ligand backbone, *i.e.* [(HC(C(*t*-Bu)N(2,6-*i*-Pr$_2$C$_6$H$_3$)$_2$)AlMe][B(C$_6$F$_5$)$_4$] proved to be an effective catalyst for hydrogenation of several imines (Scheme 5.14). In these reductions, 10 mol% catalyst was employed at temperatures between 20 and 80 °C with 1.5–6 bar of dihydrogen pressure, achieving quantitative yields in 3–200 h. Mechanistic studies revealed an autocatalytic mechanism. As the product amine increases in concentration, it takes over the role of the base in the FLP activation of dihydrogen. The generated ammonium cation transfers a proton to the imine substrate, while the aluminium hydride delivers hydride to the transient iminium cation. These findings demonstrate the subtle steric demands that must be explored and addressed in the design of new FLP catalysts.

5.6 Carbocations

The 2007 observation by Bertrand and coworkers[23] that (alkyl)(amino)carbenes could react with dihydrogen in contrast to N-heterocyclic carbenes (see Section 2.11) foreshadowed the consideration of carbon-based Lewis acids in FLP chemistry. Arduengo and coworkers[24] showed that the combination of the N-heterocyclic carbene C$_3$H$_2$(N*t*-Bu)$_2$ and the Lewis acidic trityl cation salt [Ph$_3$C][BF$_4$] under dihydrogen resulted in activation of dihydrogen, affording the imidazolium salt [C$_3$H$_2$(N*t*-Bu)$_2$H][BF$_4$] and triphenyl-methane, Ph$_3$CH (Scheme 5.15). This observation demonstrated a unique FLP in which both the Lewis acid and base are carbon-based.

While the above chemistry might invoke a conclusion that the formation of a relatively strong C–H bond would preclude applications of carbon-based Lewis acids in catalysis, this would however prove erroneous. Clark and Ingleson[25] prepared a species formulated as an acridine-derived borenium salt [C$_{13}$H$_9$NBCl$_2$][AlCl$_4$]. This species proved to be Lewis acidic at carbon as, in combination with tri-*t*-butyl pyridine, this FLP reacted with dihydrogen to affect the addition of hydride to the central carbon of the acridine ligand with concurrent protonation of pyridine (Scheme 5.16). This finding demonstrates that despite the formal charge on boron, this cation acts as a carbon-based Lewis acid.

Building on this finding, Clark and Ingleson[26] realized that the Lewis acidity of acridine could be readily achieved by the simple formation of a *N*-methyl acridinium salt, [C$_{13}$H$_9$NMe][B(3,5-Cl$_2$C$_6$H$_3$)$_4$]. A combination of this Lewis acidic species with 2,6-lutidine

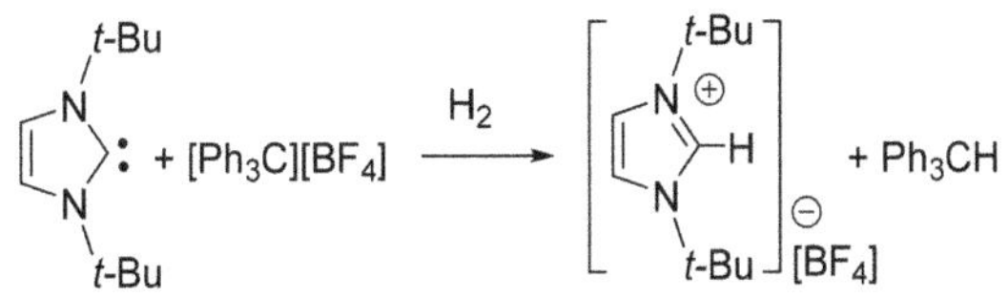

Scheme 5.15 Reaction of NHC and trityl-cation as FLP.

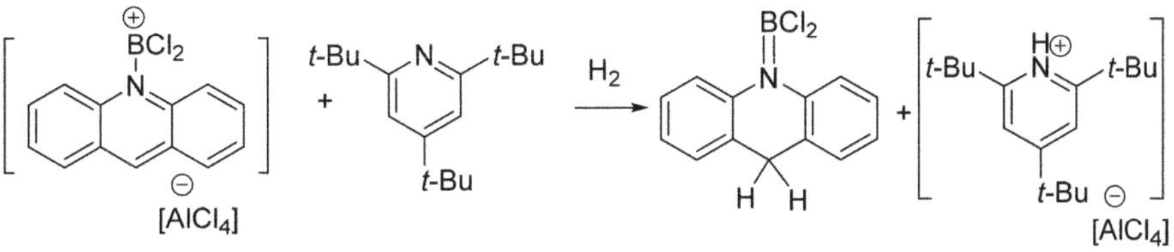

Scheme 5.16 Reactions of borenium cation, base, and dihydrogen.

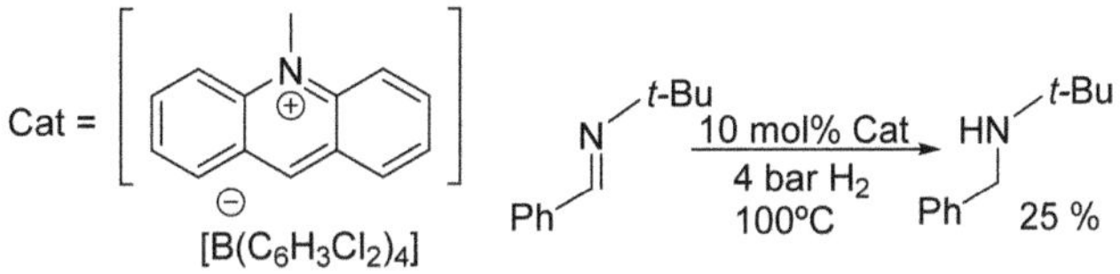

Scheme 5.17 Hydrogenation of imine with carbocation-based Lewis acid.

generated an FLP. Exposure to dihydrogen at 100 °C resulted in the formation of the salt $[2,6\text{-}Me_2C_5H_3NH][B(3,5\text{-}Cl_2C_6H_3)_4]$ and $C_{13}H_{10}NMe$, resulting from hydride delivery to the central carbon of acridinium and protonation of the base. Subsequent efforts to employ 10 mol% of this Lewis acid in the hydrogenation of PhCH=Nt-Bu using 4 bar of dihydrogen revealed a very slow reaction, achieving only 25% conversion in 72 h (Scheme 5.17). This was attributed to poor reaction kinetics for the activation of dihydrogen. This view was supported when imine reduction kinetics improved with the use of Me_2NHBH_3 as a transfer source of dihydrogen (see Section 6.3).

Another approach to a carbon-based Lewis acid was uncovered with the study of the complex $[((Ph_2PC_6H_4)_2B(\eta^6\text{-}Ph))RuCl][B(C_6F_5)_4]$.[27] Reactions with sterically unencumbered phosphines resulted in the formation of donor–acceptor adducts of the form $[((Ph_2PC_6H_4)_2B(\eta^5\text{-}C_6H_5\text{-}o\text{-}PR_3))RuCl][B(C_6F_5)_4]$, *via* formation of a C–P bond with the π-arene moiety, despite the presence of a tricoordinate boron center. Frustrating this dative interaction with the use of Mes_3P generated an FLP, as no coordination was observed. However, exposure of this FLP to dihydrogen resulted in heterolytic cleavage of dihydrogen affording *ortho*- and *para*-substituted isomers of $[((Ph_2PC_6H_4)_2 B(\eta^5\text{-}C_6H_6))RuCl]$ together with $[Mes_3PH][HB(C_6F_5)_3]$ (Scheme 5.18). Thus, the net effect is to add hydride to the arene fragment of the ruthenium complex and protonate the phosphine. Extending this to catalysis, 1–10 mol% of this novel carbon-based Lewis acid was

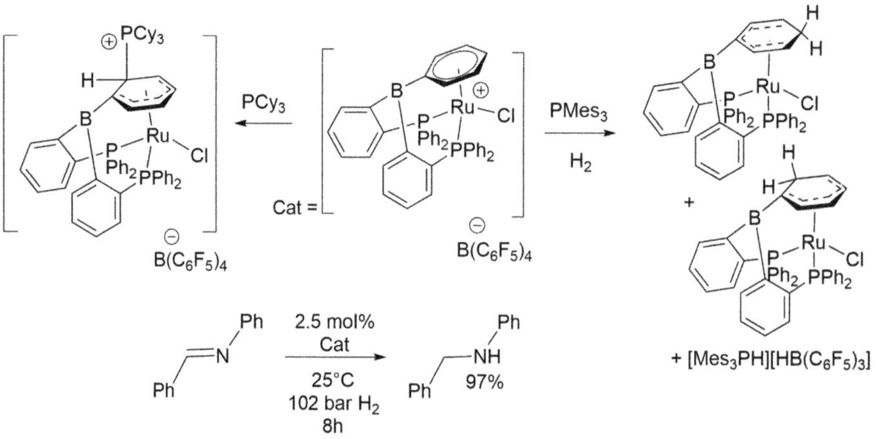

Scheme 5.18   Reactivity of Ru complex as carbon-based Lewis acid.

shown to affect the quantitative FLP reduction of imines under 102 bar of dihydrogen pressure at room temperature in 2–8 hours. It is noteworthy that this system is a curious example in which a metal center is ancillary to the site of reactivity on the carbon-based ligand.

5.7   Carbon Lewis Bases

In 2010, Piers and coworkers[28] reported the synthesis of the anti-aromatic borole $(C_6F_5C)_4B(C_6F_5)$. This extremely electron-deficient species was shown to react with dihydrogen, affording the dihydroborole $(C_6F_5CCH(C_6F_5))_2B(C_6F_5)$ (Scheme 5.19). The proposed mechanism involves the addition of dihydrogen across a B–C bond, affording a transient borane-substituted butadiene, which undergoes rapid ring closure with hydride migration affording a mixture of the *cis* and *trans* isomers of the product dihydroborole. Mechanistically, this is analogous to the FLP activation of dihydrogen by the species $R_2PB(C_6F_5)_2$ (see Section 2.10).

Also in 2010, the Alcarazo group[29] extended FLP chemistry to the intermolecular FLP derived from the carbon-based donor, $(Ph_3P)_2C$ and $B(C_6F_5)_3$. This carbone/borane combination was shown to react at room temperature and 1 bar of dihydrogen to give the salt $[(Ph_3P)_2CH][HB(C_6F_5)_3]$. A subsequent computational study[30] examined this reaction as well as the related heavier congener ylidones, finding that the dihydrogen activation strongly depends on the nature of both the central atom and the donors, as lower barriers are associated with early transition states. Moreover, these calculations suggest that the heavier analogs are more likely to undergo a facile reaction with dihydrogen.

Krempner and coworkers[31] developed a sterically protected zwitterion containing a carbanionic center. This species, $[C(SiMe_2OCH_2CH_2OMe)_3Na]$, in combination with several boranes, was shown to activate dihydrogen. Of particular interest, the strongly basic carbanion permits the activation of dihydrogen with weakly Lewis acidic boranes such as BPh_3, $HBMes_2$, and BEt_3 (Scheme 5.20). Moreover, structural data for the product salts,

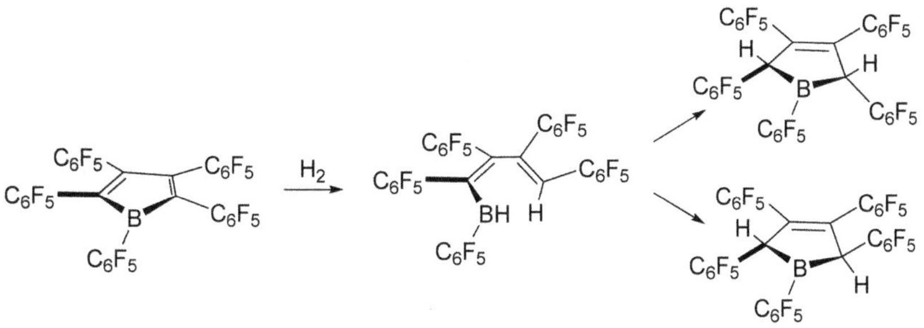

Scheme 5.19 Reactions of Piers' borole with dihydrogen.

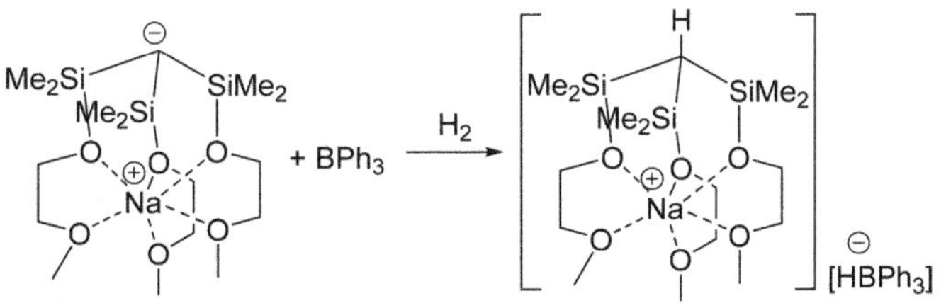

Scheme 5.20 Reaction of carbanion, borane, and dihydrogen.

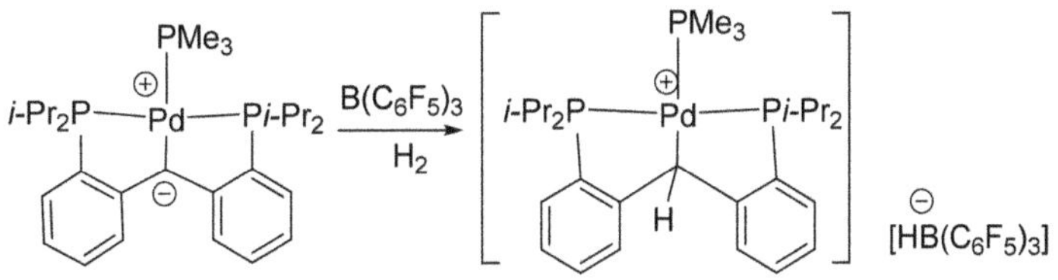

Scheme 5.21 FLP activation of dihydrogen by Pd-carbene/B(C$_6$F$_5$)$_3$.

[HC(SiMe$_2$OCH$_2$CH$_2$OMe)$_3$Na][HBR$_3$] (R$_3$ = Ph$_3$, HMes$_2$) revealed the close approach of the C–H and H–B fragments of 2.18 and 2.47 Å, respectively. This unique behavior with weak Lewis acids prompted Krempner to label these systems 'inverse' FLPs.

A study by the Iluc group[32] showed that the carbene bound to Pd complex (*i*-Pr$_2$PC$_6$H$_4$)$_2$CPd(PMe$_3$) reacts in the presence of B(C$_6$F$_5$)$_3$ and dihydrogen in an FLP fashion, affording [[(*i*-Pr$_2$PC$_6$H$_4$)$_2$C(H)Pd(PMe$_3$)][HB(C$_6$F$_5$)$_3$] (Scheme 5.21). Here again, the metal center is thought to be ancillary to the ligand reactivity.

5.8 Silicon-based Lewis Acids

In 2011, the Müller group[33] probed the reaction of the FLP derived from the combination of silylium cation [(C$_6$Me$_5$)$_3$Si][B(C$_6$F$_5$)$_4$] and PMes$_3$, demonstrating its ability to activate dihydrogen (Scheme 5.22) with generation of the corresponding silane and the phosphonium salt. Efforts to extend such reactivity to less sterically hindered

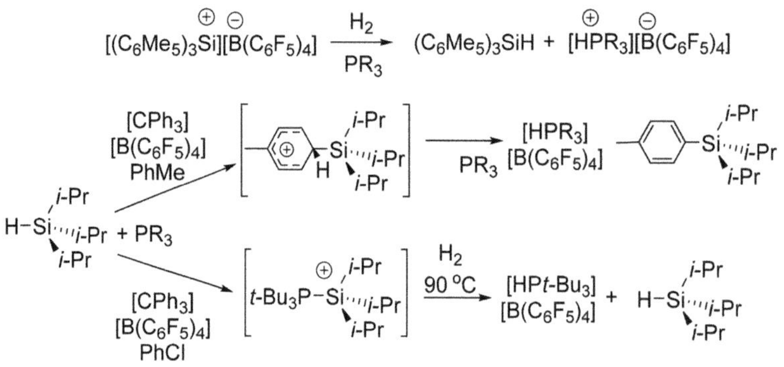

Scheme 5.22 Generation and reactivity of silyl-cations in differing solvents.

trialkylsilylium cations were found to be more complex.[34] In aromatic solvents, interactions of the silyl cation with toluene or benzene generates a Brønsted acid, which is readily deprotonated to give the aryl-silane. Such arene reactions were circumvented using chlorobenzene, providing an avenue to classical silylium–phosphine adducts. Ashley *et al.*[35] later showed that these adducts also exhibited FLP behavior, activating dihydrogen at elevated temperatures, generating silane and the phosphonium salt (Scheme 5.22).

5.9 Silylene Donors

The Müller group probed the reaction of the FLP derived from the combination of silylium cation $[(C_6Me_5)_3Si][B(C_6F_5)_4]$ and silylenes.[34,36] The ability to activate dihydrogen established the first fully silicon based FLP. While the Lewis acid accepts hydride to generate the corresponding silane, the protonation of the silylene prompts isomerization to a hydrogen-bridged disilyl cation (Scheme 5.23).

In a subsequent study,[37] the dialkylsilylene was found to react with dihydrogen in the presence of a small amount of Lewis acid (BPh₃, BEt₃) or a base (PPh₃, PEt₃, NPh₃, NEt₃), even at low temperatures, to generate the corresponding dihydrosilane. These observations were rationalized by an FLP mechanism in which the silylene acts as either acidic or basic components of an FLP.

In 2019 Driess and coworkers[38] described the first Si(0) species stabilized by bis(N-heterocyclic silylenes). In the presence of the Lewis acid BPh₃, this silylone affects the heterolytic cleavage of dihydrogen (1 bar) affording the corresponding hydridosilyliumylidene hydroborate salt. This extension of FLP chemistry to exceptionally electron-rich Si(0) species in combination with the relatively weak acid BPh₃ is another example in which there appears to be a threshold of combined Lewis acidity and basicity necessary for dihydrogen activation (Scheme 5.24). In a related effort, a collaboration with the Lips group[39] described the reaction of the NHC-coordinated trisilacyclopropylidene $(NHC)Si(Mes_2SiSiMes_2)$ and the Lewis acid $B(C_6F_4H)_3$ with dihydrogen at room temperature. This proceeded to give the salt $[(NHC)SiH(Mes_2SiSiMes_2)][HB(C_6F_4H)_3]$ (Scheme 5.24).

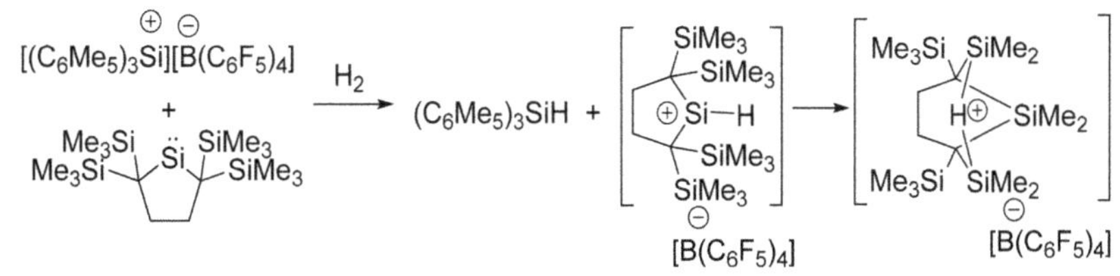

Scheme 5.23 Dihydrogen activation by [(C₆Me₅)₃Si][B(C₆F₅)₄] and silylene.

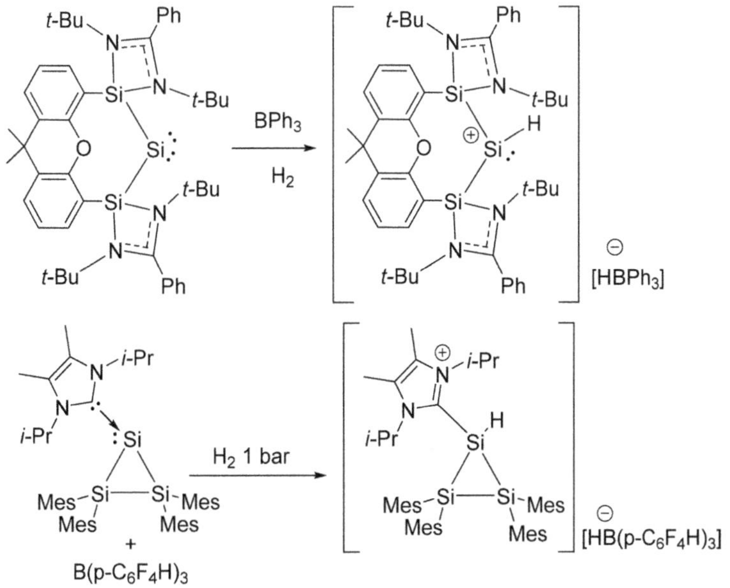

Scheme 5.24 Reactions of low valent silicon species with borane and dihydrogen.

5.10 Tin-based Lewis Acids

Seeking a readily accessible Lewis acid, Ashley *et al.*[40] examined the use of (PhCH₂)₃Sn(OTf) in FLP hydrogenations. These researchers showed that this species could activate dihydrogen and mediate the stoichiometric reduction of imines. However, in the catalytic reaction, this Lewis acid only gave low yields, as the tin cation was deactivated by competing substituent scrambling processes.

However, in subsequent work, Ashley and coworkers[41] reckoned that the Lewis acidity of the Sn center would be increased with larger substituents and this could impart the ability to activate dihydrogen. Indeed, the Sn-based Lewis acid, *i*-Pr₃Sn(OTf) proved effective in mediating the catalytic hydrogenation of a series of imines, ketones, and aldehydes at 120–180 °C under 10 bar of dihydrogen pressure (Scheme 5.25). In the case of carbonyl reductions, the addition of base was required for catalysis. Mechanistically, this reaction is thought to proceed *via* the generation of the stannane, which adds across the double bond of the substrate. Subsequent proton transfer from the base to the

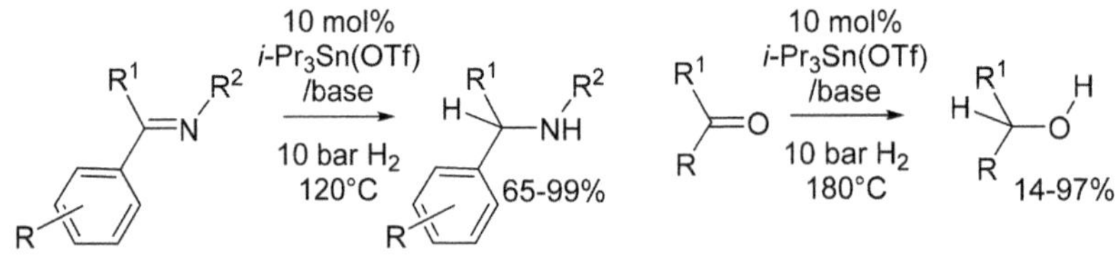

Scheme 5.25 Examples of hydrogenation catalysis using *i*-Pr$_3$Sn(OSO$_2$CF$_3$)/base as the catalyst.

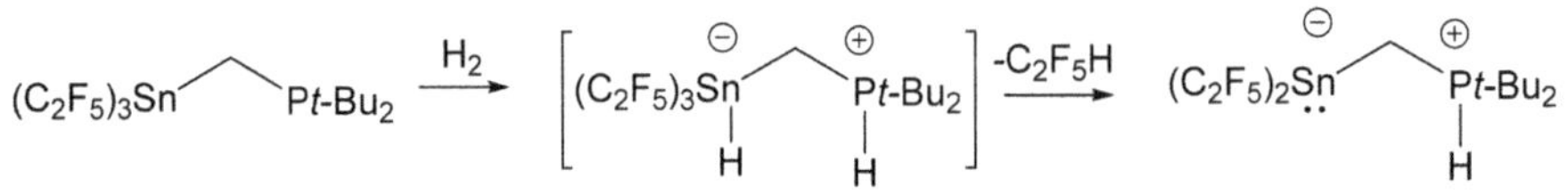

Scheme 5.26 Reaction of intramolecular Sn/P FLP with dihydrogen.

N or O of the substrate bound to tin releases the reduced product and regenerates the tin cation for further catalysis. This finding is particularly innovative as this species is simply and inexpensively prepared. Moreover, it provided a catalyst that is not quenched by moisture. Details of the activation of dihydrogen by Sn/N FLPs[42] and the hydrogenation of carbonyl groups have been examined computationally,[43,44] while a recent study has addressed the role for the triflate anion in the activation of dihydrogen.[45]

In the last year, Mitzel and coworkers[46] prepared the first intramolecular Sn/P FLP (C$_2$F$_5$)$_3$SnCH$_2$P*t*-Bu$_2$. This species reacts slowly over 6 days with dihydrogen at 11 bar, subsequently undergoing reductive elimination of C$_2$F$_5$H, giving the tin(II)/P zwitterion, (C$_2$F$_5$)$_2$SnCH$_2$P(H)*t*-Bu$_2$ (Scheme 5.26).

5.11 Phosphorus-based Lewis Acids

Recognizing that FLP reductions of N-heterocycles were possible, we began to explore the possibility of FLP reduction of triphosphabenzene. However, control experiments revealed that this molecule reacts with dihydrogen on its own! The resulting product is a 3,5-bicyclic species (Scheme 5.27).[47] Variable temperature NMR studies together with *para*-hydrogen experiment data demonstrated that dihydrogen and triphosphabenzene react *via* the initial formation of the 1,4-addition species, which undergoes subsequent rearrangement to give the observed bicyclic product. Computational studies revealed that the interaction of dihydrogen with a boat conformer of triphosphabenzene amounts to the polarization of dihydrogen between a Lewis acidic cationic phosphorus center and a basic carbanion. Thus, this reaction pathway is directly analogous to that of an FLP-dihydrogen reaction.

Further developments of P(III)-based Lewis acids are beginning to emerge. Burford and coworkers[48] have explored the FLP reactivity of [(bipy)$_3$P]$^{3+}$ trication in the presence of *t*-Bu$_3$P and dihydrogen. While the formation of [*t*-Bu$_3$PH]$^+$ suggests FLP activation of dihydrogen, the fate and nature of the hydride addition to the Lewis acidic center

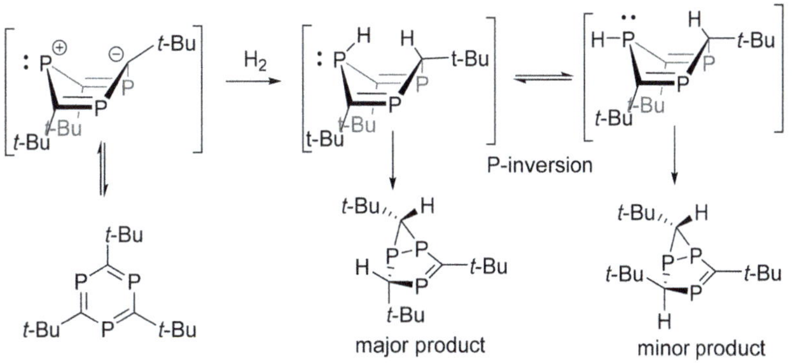

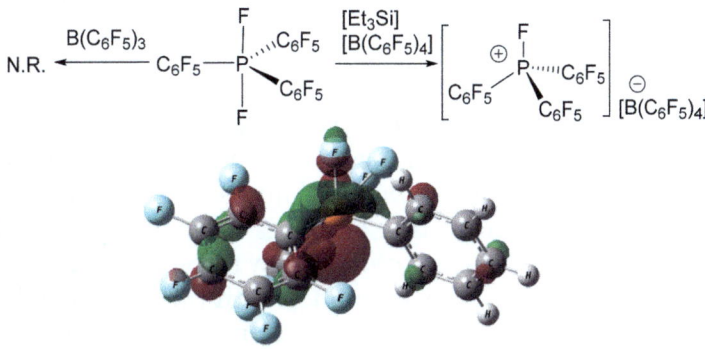

Scheme 5.27 Stoichiometric reaction of triphosphabenzene with dihydrogen.

Scheme 5.28 Synthesis and LUMO of [FP(C$_6$F$_5$)$_2$(C$_6$H$_5$)][B(C$_6$F$_5$)$_4$].

could not be identified. Nonetheless, these observations suggest that, generally, cationic main group elements have unexplored Lewis acid reactivity.

The above findings prompted consideration of P(v)-based Lewis acids. To that end, targeted electron-deficient P(v) cations were readily prepared by the oxidation of P(iii) species with XeF$_2$. Abstraction of fluoride from the resulting difluoro phosphorane generated a fluorophosphonium cation. Incorporating additional electron-withdrawing substituents afforded a series of highly electrophilic phosphonium cations (EPCs). Indeed, these species were strong Lewis acids. For example, the reaction of difluorophosphorane Ph$_2$PF$_2$(C$_6$F$_5$) with B(C$_6$F$_5$)$_3$[49] showed facile fluoride ion exchange between boron and phosphorus, inferring that the P(v)-cation and the borane had similar fluorophilicity. The incorporation of additional electron-drawing groups further enhanced the Lewis acidity of the fluorophosphonium cation. For example, efforts to remove a fluoride ion with B(C$_6$F$_5$)$_3$ from the F$_2$P(C$_6$F$_5$)$_3$ were unsuccessful, inferring the targeted cation is more fluorophilic than B(C$_6$F$_5$)$_3$.[50] It is only with the use of an exceptionally strong Lewis acid, [Et$_3$Si][B(C$_6$F$_5$)$_4$], that the EPC salts [FP(C$_6$F$_5$)$_2$R][B(C$_6$F$_5$)$_4$] (R = Ph, C$_6$F$_5$) could be isolated (Scheme 5.28).

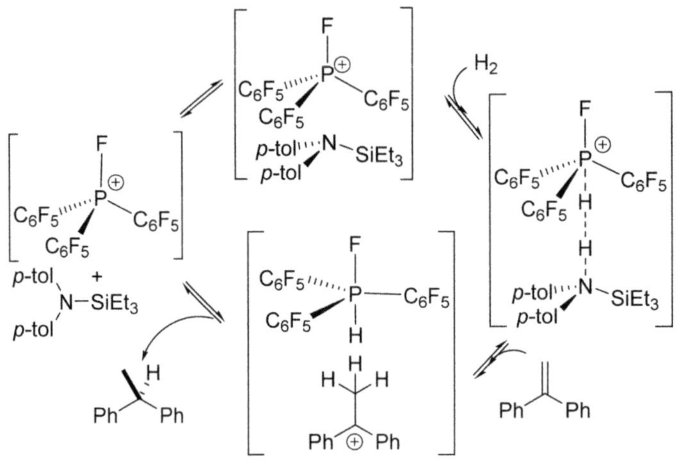

Scheme 5.29 Proposed mechanism for the hydrogenation catalysis by the FLP derived from p-tol$_2$NSiEt$_3$ and [FP(C$_6$F$_5$)$_3$][B(C$_6$F$_5$)$_4$].

It is noteworthy that these EPCs exhibit Lewis acidity that is distinct from that of boranes. While the Lewis acidity of boranes is derived from the presence of a vacant p-orbital, the Lewis acidity of fluorophosphonium cations arises from a low energy σ*-orbital. The major component of this LUMO is oriented opposite the P–F bond. Thus, these Lewis acidic sites are also umbrellaed by the arene rings (Scheme 5.28).

The species [FP(C$_6$F$_5$)$_3$][B(C$_6$F$_5$)$_4$] was combined with the sterically demanding base p-Tol$_2$NSiEt$_3$, generating an FLP.[51] Under 4 bar of dihydrogen and in the presence of 1 mol% of the Lewis acid and 20 mol% of the base, almost quantitative hydrogenation of aryl-olefins and silylenol ethers was achieved (Scheme 5.29). The proposed mechanism is directly analogous to that proposed for P/B FLPs with the approach of the nitrogen base and the P-Lewis acid, generating an encounter complex that reacts with dihydrogen (Scheme 5.29). Nonetheless, these reductions were slower than those mediated by B(C$_6$F$_5$)$_3$.

5.12 Early Transition Metal Acids

Before we discovered FLP chemistry, we reported cations of the form [CpTi(NPR$_3$)Me(PR$_3$)][MeB(C$_6$F$_5$)$_3$][52] could be formed with small phosphines, but bulky phosphines such as PR$_3$ (R = C$_6$H$_2$Me$_3$, t-Bu) did not coordinate to the titanium atom. While this finding led us to investigate the independent reactions of bulky phosphines and B(C$_6$F$_5$)$_3$, this observation also foreshadowed the generation of early transition metal based FLPs. However, Wass $et\ al.$ first reported on FLPs incorporating early metal Lewis acids in 2011. These authors described the synthesis of the complex [Cp*$_2$Zr(ClC$_6$H$_5$)OC$_6$H$_4$Pt-Bu$_2$][B(C$_6$F$_5$)$_4$],[53] which reacted with dihydrogen, providing the Zr-hydride complex with a pendant phosphonium cation, [Cp*$_2$Zr(H)OC$_6$H$_4$PHt-Bu$_2$][B(C$_6$F$_5$)$_4$] (Scheme 5.30), affirming its behavior as a Zr/P based FLP.

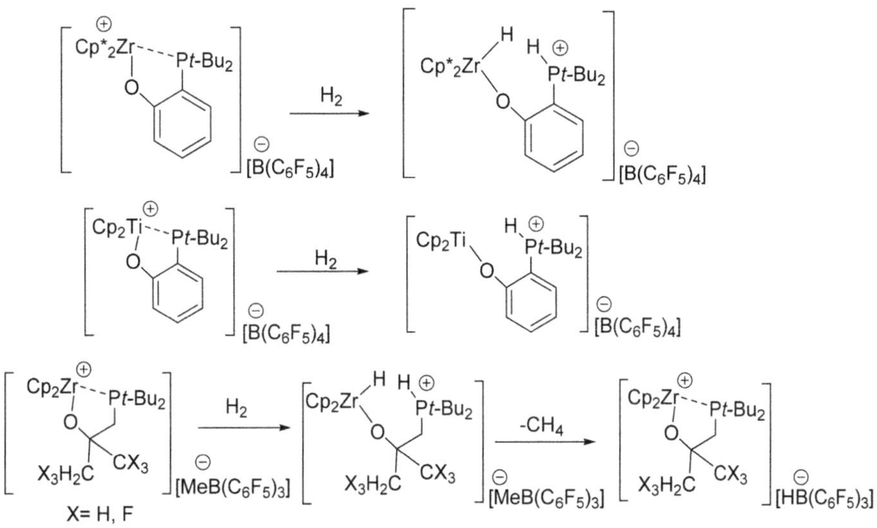

Scheme 5.30 Further examples of transition metal FLPs.

In a subsequent report, Wass and coworkers[54] probed the issue of steric frustration in the analogous species $[Cp_2ZrOC_6H_4Pt\text{-}Bu_2][B(C_6F_5)_4]$ where the phosphorus is bound to Zr. This species proved unreactive with dihydrogen, affirming that steric frustration between the Lewis acidic Zr and the basic phosphine is required. In contrast, the titanium analog $[Cp_2TiOC_6H_4Pt\text{-}Bu_2][B(C_6F_5)_4]$ does react with dihydrogen, however the transient titanium-hydride loses a hydrogen atom, affording the Ti(III) species $[Cp_2TiOC_6H_4PHt\text{-}Bu_2][B(C_6F_5)_4]$.[55] Similarly, modification of the phosphine substituents, as in $[Cp^*_2ZrOC_6H_4Pi\text{-}Pr_2][B(C_6F_5)_4]$, resulted in an incomplete reaction with dihydrogen, inferring that while dissociation of phosphine from Zr permits reaction with dihydrogen, the reverse reaction in which dihydrogen is evolved from $[Cp^*_2Zr(H)OC_6H_4PHi\text{-}Pr_2][B(C_6F_5)_4]$ is also possible. The other structural perturbations $[Cp_2Zr(O(CX_3)_2CH_2t\text{-}PBu_2)][MeB(C_6F_5)_3]$ (X = H, F) also reacted with dihydrogen. Protonation of the methyl borate and hydride abstraction, led to the formation of the product, $[Cp_2Zr(O(CX_3)_2CH_2Pt\text{-}Bu_2][HB(C_6F_5)_3]$ (Scheme 5.30).[56]

A collaborative effort from the Wass and Manners groups[57] showed that the Zr cation $[Cp^*_2Zr(OMes)][B(C_6F_5)_4]$ reacted with D_2 gas in the presence of $t\text{-}BuN\text{=}CHPh$ to effect the heterolytic cleavage of D_2, affording $Cp^*_2ZrD(OMes)$ and $[t\text{-}BuND\text{=}CHPh][B(C_6F_5)_4]$. Moreover, using 10 mol% of the indenyl-derived cation $[Ind_2Zr(OMes)]^+$, bulky imines could be catalytically hydrogenated using 1.5 bar at 25 °C in 90 minutes (Scheme 5.31). Sterically unencumbered imines were not reduced, presumably a result of the formation of an adduct with the Lewis acidic Zr cation. The proposed mechanism was an FLP process, involving activation of dihydrogen by the imine and Zr-Lewis acid, followed by hydride delivery from the Zr to the imine carbon, and release of the amine from Zr regenerating the catalyst.

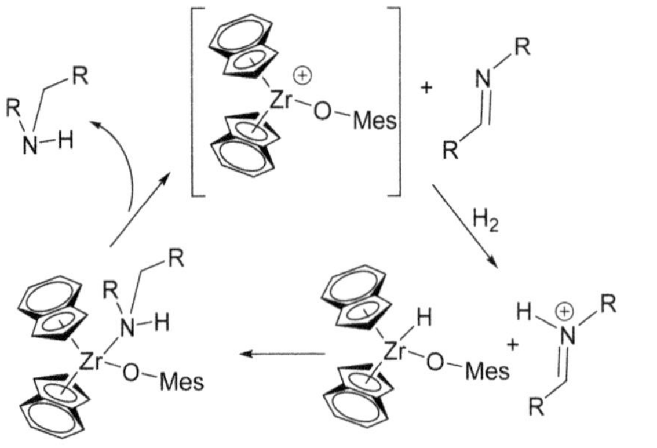

Scheme 5.31 Catalytic imine reduction by Zr-cation.

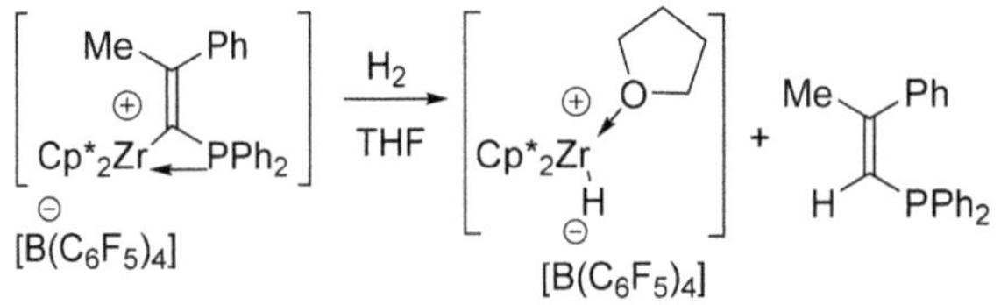

Scheme 5.32 Reaction of [Cp$_2$ZrC(=C(Me)Ph)PPh$_2$][B(C$_6$F$_5$)$_4$] with dihydrogen.

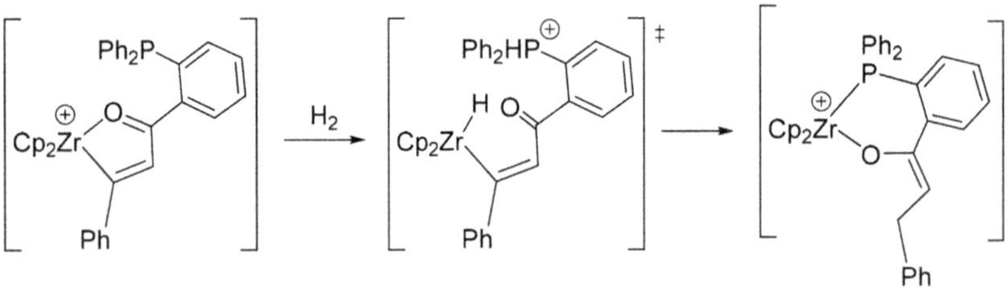

Scheme 5.33 Reaction of Zr/P FLP [Cp$_2$ZrC(Ph)=CHC(O)C$_6$H$_4$PPh$_2$][B(C$_6$F$_5$)$_4$] with dihydrogen.

In a closely related system, the Erker group[58] reported the FLP reaction of [Cp*$_2$Zr(O(CH$_2$)$_2$Ni-Pr$_2$)][B(C$_6$F$_5$)$_4$] with dihydrogen, generating [Cp*$_2$ZrH(O(CH$_2$)$_2$NHi-Pr$_2$)][B(C$_6$F$_5$)$_4$]. 1–4 mol% of this Zr/N FLP also acted as an effective catalyst for the reduction of a series of alkynes, olefins dienes, and enamines using only 1.5 bar pressure of dihydrogen. The Erker group also probed the Zr/P based FLP, [Cp$_2$ZrC(=C(Me)Ph)PPh$_2$][B(C$_6$F$_5$)$_4$][59] obtained by reaction of methylzirconocene cations with diphenylphosphino-substituted acetylene. While this species behaved as an FLP with other small molecules, reaction with dihydrogen cleaved the Zr–C bond affording [Cp$_2$ZrH(THF)][B(C$_6$F$_5$)$_4$] and MeC(Ph)C(H)PPh$_2$ (Scheme 5.32).

In another perturbation of a Zr/P FLP, the Erker group[60] found that [Cp$_2$ZrC(Ph)=CH-C(O)C$_6$H$_4$PPh$_2$][B(C$_6$F$_5$)$_4$] reacted with dihydrogen generating an acidic phosphonium cation which prompted a rearrangement affording the species [Cp$_2$ZrOC(C$_6$H$_4$PPh$_2$)(=CH(CH$_2$Ph)][B(C$_6$F$_5$)$_4$] (Scheme 5.33).

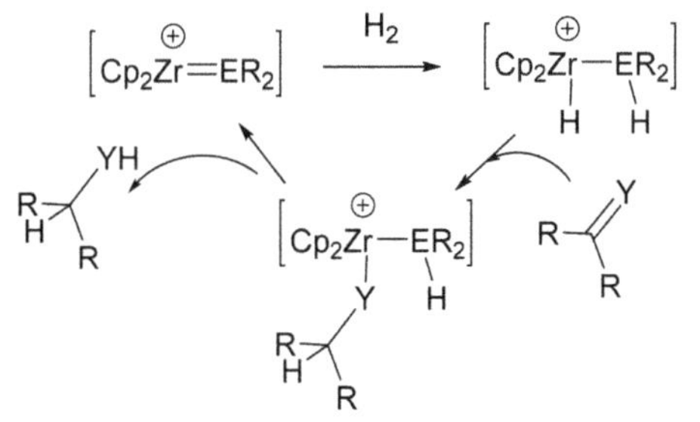

Scheme 5.34 Catalytic hydrogenation mediated by [Cp$_2$Zr=ER$_2$]X.

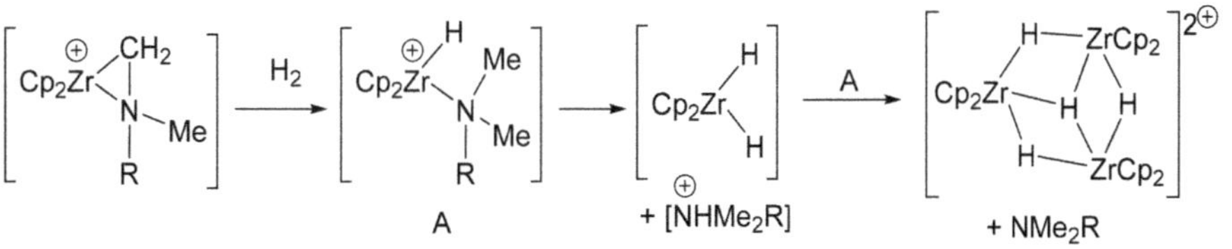

Scheme 5.35 Reaction of zirconaziridinium complex with dihydrogen.

The zirconocene cations of the form [Cp$_2$Zr=ER$_2$]X (E = N, P; X = MeB(C$_6$F$_5$)$_3$ or B(C$_6$F$_5$)$_4$)[61] also behaved as an intramolecular FLP despite the direct Zr–E bonds (Scheme 5.34). These species serve as catalysts for the hydrogenation of imines, enamines, olefins, and alkynes *via* the addition of dihydrogen across the Zr=E bonds, insertion of the substrate into the Zr–H bond, and subsequent protonolysis regenerating the catalytically active Zr/P FLP.

Most recently, Budzelaar and coworkers[62] have probed the reaction of the zirconaziridinium complex [Cp$_2$Zr(CH$_2$NMeR)][B(C$_6$F$_5$)$_4$] with dihydrogen. While the initial reaction involved a fast sigma-bond metathesis of the Zr–C bond, the resulting zirconocene amino hydride complexes behaved as FLPs, prompting the heterolytic splitting of dihydrogen affording ammonium salt and a trimetallic zirconium hydride [(Cp$_2$Zr(μ^2-H))$_3$(μ^3-H)][B(C$_6$F$_5$)$_4$]$_2$ (Scheme 5.35).

5.13 Late Transition Metal Acids

The behavior of transition metals as Lewis acids is not limited to the early metals. Another system that activated dihydrogen exploiting an acidic metal and a donor involved the reaction of [Ni(Et$_2$P(CH$_2$)$_3$PEt$_2$)$_2$]$^{2+}$ and NEt$_3$ with dihydrogen.[63] This reaction results in the formation of [HNi(Et$_2$P(CH$_2$)$_3$PEt$_2$)$_2$]$^+$ and [HNEt$_3$]$^+$. While this reaction was reported in 2002, before the articulation of the concept of FLPs, it is clear in retrospect that mechanistically this involves the interaction of dihydrogen with the Lewis acidic Ni-dication, prompting protonation of the amine and the net delivery of a hydride to the nickel center.

In a related system, Dubois, Bullock and coworkers prepared the Ni(II) complexes [((RN(CH$_2$)$_2$PR)$_2$)$_2$Ni]$^{2+}$.[64] Reaction with dihydrogen resulted in the protonation of the pendant amine groups and hydride delivery to the Ni center. The computed mechanism was proposed to proceed *via* coordination of dihydrogen to Ni, with subsequent proton transfer to nitrogen generating a transient Ni(0) species which was electrochemically reduced to transient Ni(II) species with the net release of two protons (Scheme 5.36). The authors noted the direct analogy to FLPs, moreover, as well as the relation to the active site of the enzyme, hydrogenase (see Section 6.12). In subsequent work, related Fe and Mn species have been studied and shown to behave in a similar fashion providing new electro-catalysts for dihydrogen oxidations.[65]

Bullock and coworkers[66] have also examined the related activation of dihydrogen by the Mo-complex [((RN(CH$_2$)$_2$PR)$_2$)$_2$MoCp(CO)]$^+$. Heterolytic cleavage of dihydrogen resulting in the protonation of a pendant amine and hydride delivery to Mo was observed. The proton and hydride underwent rapid exchange, although the inclusion of a more basic amine or a more electron-donating phosphine slowed such exchange.

In a 2015 report Renaud and coworkers[67] described the reaction of the complex (C$_5$Ph$_2$(NPhCH$_2$)O)Fe(CO)$_3$ with Me$_3$NO. This generated the transient transition-metal based FLP, where the vacant site on Fe acted as a Lewis acid in conjunction with the pendant basic oxygen atom. 5 mol% of this species acted a catalyst for the condensation of a series of ketone and amine combinations, and in the presence of dihydrogen mediated the reduction of generated imine affording the corresponding amine product (Scheme 5.37). Computational results support the notion of dihydrogen activation *via* the Fe/O FLP. The structurally related Fe species,[68] [((CH$_2$NMe)$_2$C$_5$Ph$_2$NH*i*-Pr)Fe(CO)$_2$][B(C$_6$F$_5$)$_4$] was also generated in a similar reaction of the Fe-tricarbonyl cation with Me$_3$NO. This species acted as an Fe/N FLP and mediated the analogous hydrogenations of *in situ* generated imines from the combination of aldehydes and amines using just 2 mol% catalyst at room temperature.

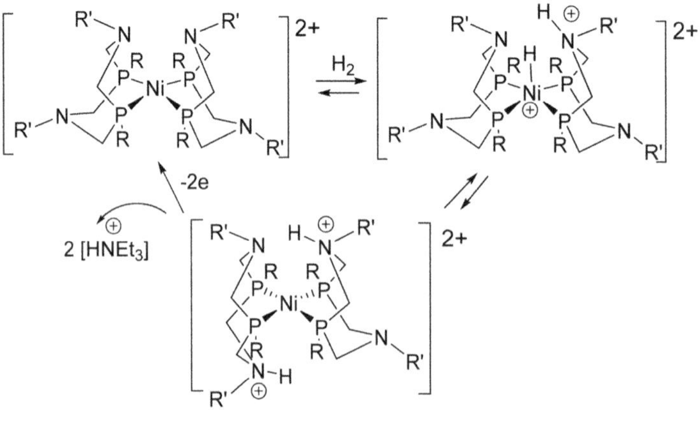

Scheme 5.36 Electrocatalytic oxidation of dihydrogen.

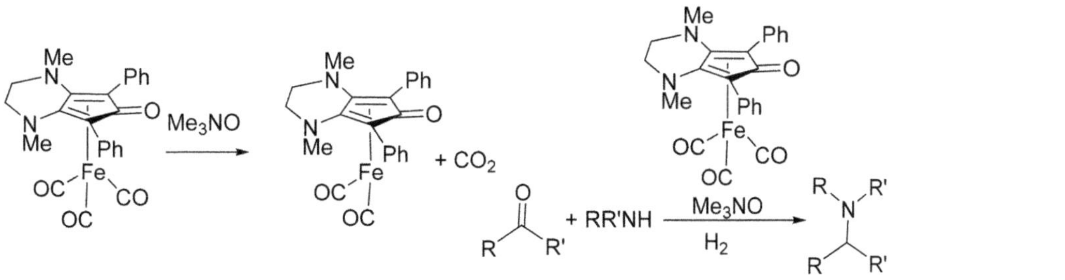

Scheme 5.37 Generation of the Fe-cation/N intramolecular FLP and its use in catalysis.

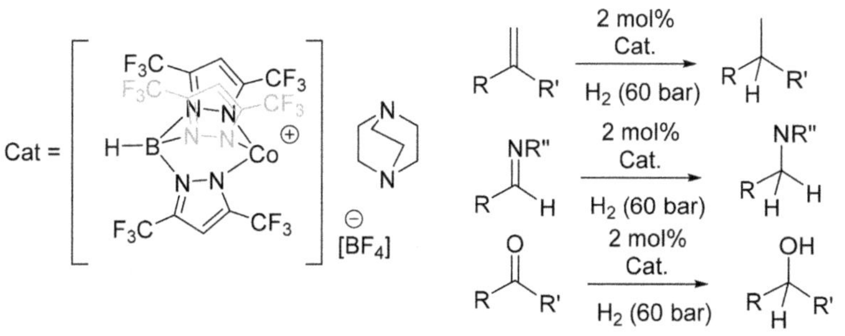

Scheme 5.38 Catalytic hydrogenation by a Co-cation/N FLP.

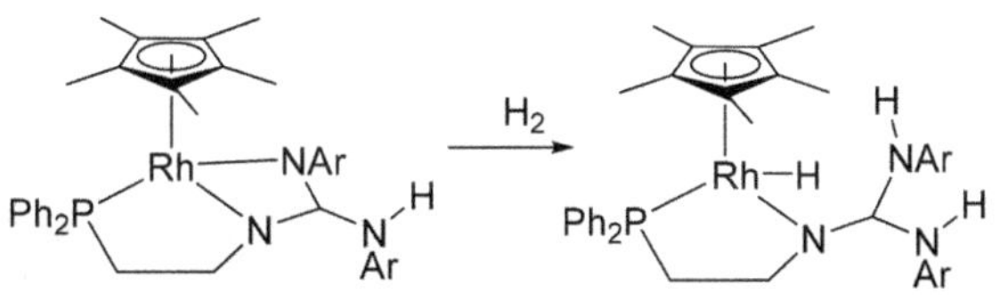

Scheme 5.39 Reaction of a Rh/N 'dormant' FLP with dihydrogen.

The tris(pyrazolyl)borate cobalt cation,[69] $[HB(N_2C_3(CF_3)_2)_3Co]^+$, in the presence of a Lewis base, also acted as an FLP. It catalyzed the homogeneous hydrogenation of a range of substrates including carbonyls, alkenes, enamines, and imines using 60 bar of dihydrogen at 60 °C (Scheme 5.38). Mechanistic studies affirmed an FLP process. In a conceptually related Ir(III)-system,[70] $[HB(N_2C_3Me_2)_3IrPh(C_5H_3NR)]^+$ (R = Me, Ph), the inclusion of the pendant pyridine fragment generated an intramolecular FLP that also reacted with dihydrogen, affording $[HB(N_2C_3Me_2)_3IrHPh(C_5H_3N(H)R)]^+$.

In addition, a recent example demonstrated the reaction of dihydrogen with the guanidine-derived phosphine-Rh complex $[Cp*Rh(Ph_2PCH_2CH_2NC(Np\text{-}tol)(NHp\text{-}tol))]$ $[SbF_6]$,[71] affording the product containing a Rh-hydride and a protonated pendant nitrogen, illustrating that the initial complex acted as a 'dormant' FLP (Scheme 5.39).

5.14   Transition Metal Bases

A question arising from the previous sections is: can an electron-rich metal center play the opposing role and thus act as a Lewis base in concert with a Lewis acid to generate an FLP? One relatively early example of this notion was reported by Erker and coworkers who prepared the Zr(II) complex $Cp_2Zr(Me_3SiCCB(C_6F_5)_2)$.[72] This species reacts with dihydrogen to afford the complex $Cp_2ZrH(Me_3SiCCBH(C_6F_5)_2)$ (Scheme 5.40). This product exhibited a close approach of the ZrH to the alkynyl carbon and of the BH to Zr. This reactivity can be viewed as derived from the FLP activation of dihydrogen by the basic Zr(II) center in concert with the pendant Lewis acidic boron. An alternative pathway involving oxidative addition of dihydrogen to Zr was considered, however computations of the possible transition states revealed that the reaction is most reasonably described as involving a Zr(II)/B FLP pathway.

Another example of a transition metal-derived basic component was described by Peters and coworkers.[73] These researchers prepared the Ni(0) complex, $PhB(C_6H_4PPh_2)_2Ni$. In this species the electron-rich metal and electron-deficient boron act in concert to activate dihydrogen to provide the product $PhBH(C_6H_4PPh_2)_2NiH$. The net result was the delivery of a proton to the Ni center and hydride to boron. Thus, this can be viewed as a Ni/B FLP (Scheme 5.41). This reaction proved reversible, as exposure of this Ni(0) species to dihydrogen and D_2 resulted in isotope scrambling and the observation of HD gas in the mixture. This reversible dihydrogen activation also allowed 1 mol% of this Ni(0) complex to mediate the catalytic hydrogenation of styrene at room temperature affording ethylbenzene quantitatively. An analogous experiment using D_2 and norbornene confirmed that the addition occurred in a *syn* fashion.

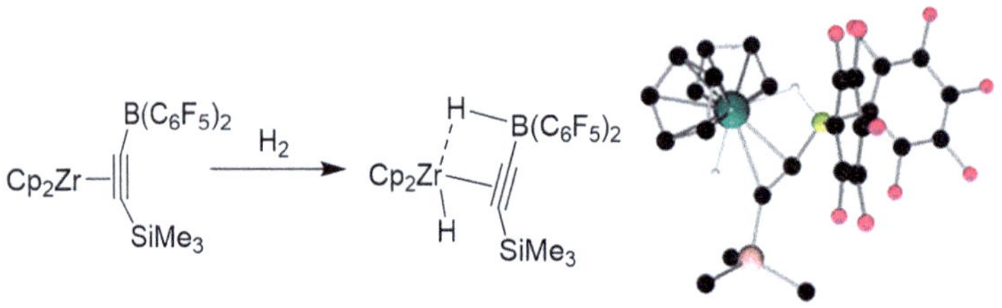

Scheme 5.40   The reaction of Zr(II)/B species with dihydrogen.

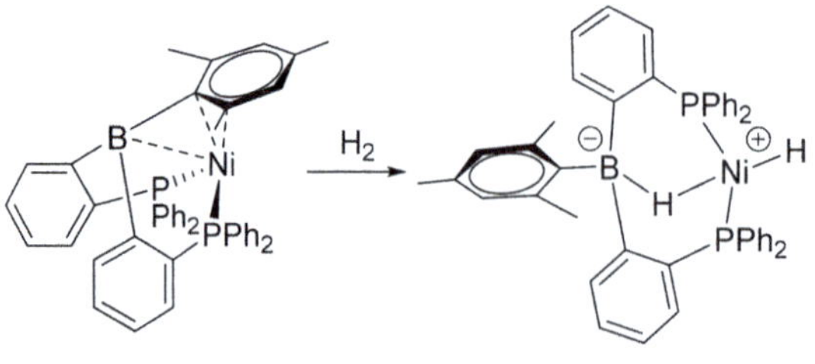

Scheme 5.41   Activation of dihydrogen by a Ni(0)/B system.

In 2015, Wass and coworkers[74] described the combination of the three-coordinate complex $(CH_2CH_2Pt\text{-}Bu_2)_2PtCO$ and $B(C_6F_5)_3$. Reaction with dihydrogen afforded the salt $[(CH_2CH_2Pt\text{-}Bu_2)_2PtHCO][HB(C_6F_5)_3]$ (Scheme 5.42). While a mechanism involving oxidative addition of dihydrogen to the Pt(0) center was considered, computational data favored an FLP type reaction pathway in which dihydrogen is activated between platinum and boron and the proposed intermediate is almost isoenergetic with dative interactions between $CH_2(CH_2Pt\text{-}Bu_2)_2PtCO$ and $B(C_6F_5)_3$. The analogous reaction of the Pd species proceeds slightly differently affording $[((CH_2CH_2Pt\text{-}Bu_2)_2Pd)_2(\mu\text{-}H)(\mu\text{-}CO)]$ $[(C_6F_5)_3BC(O)HB(C_6F_5)_3]$.[75]

Figueroa and coworkers[76] described a related intramolecular Pt/B FLP, $(C_6H_3(C_6H_3 i\text{-}Pr_2)_2NC)Pt(NCBCy_2)C_6H_3(C_6H_3 i\text{-}Pr_2)_2$ derived from the hydroboration of the bis-isocyanide complex $(C_6H_3(C_6H_3 i\text{-}Pr_2)_2NC)_2Pt$ with $HBCy_2$. This FLP species reacted with dihydrogen delivering a proton to the Pt and hydride to the pendant boron, affording the species $(C_6H_3(C_6H_3 i\text{-}Pr_2)_2NC)PtH(NCBHCy_2)C_6H_3(C_6H_3 i\text{-}Pr_2)_2$ (Scheme 5.43).

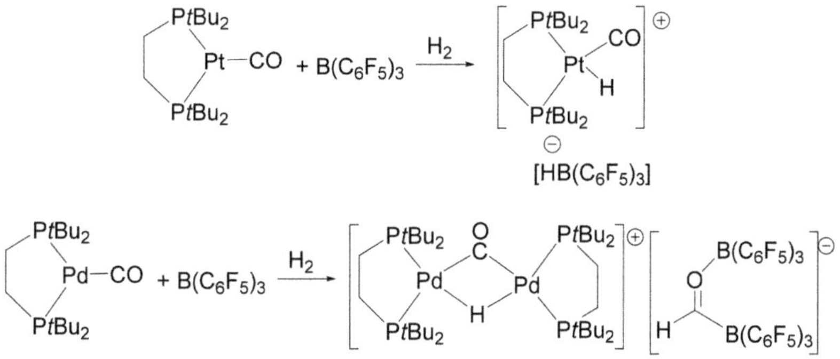

Scheme 5.42 Activation of dihydrogen by a M(0) (M = Pd, Pt)/B system.

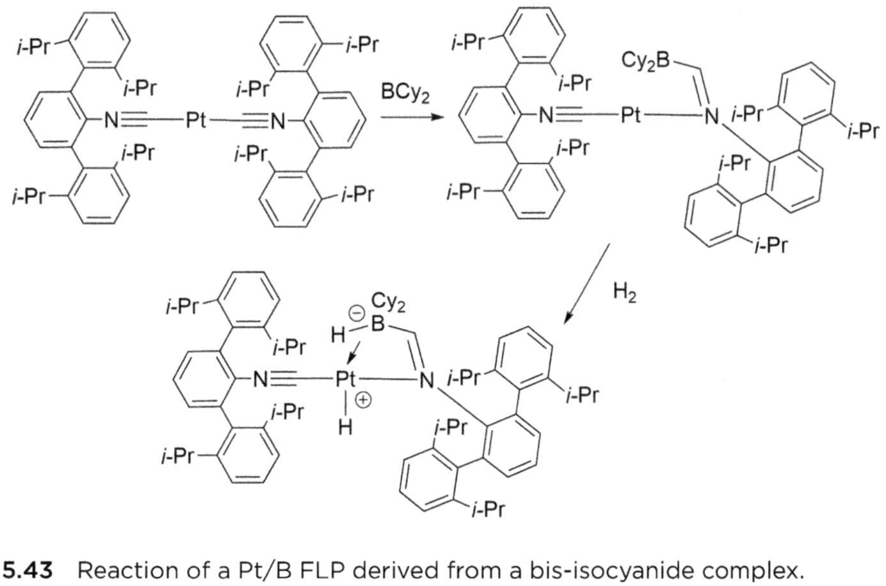

Scheme 5.43 Reaction of a Pt/B FLP derived from a bis-isocyanide complex.

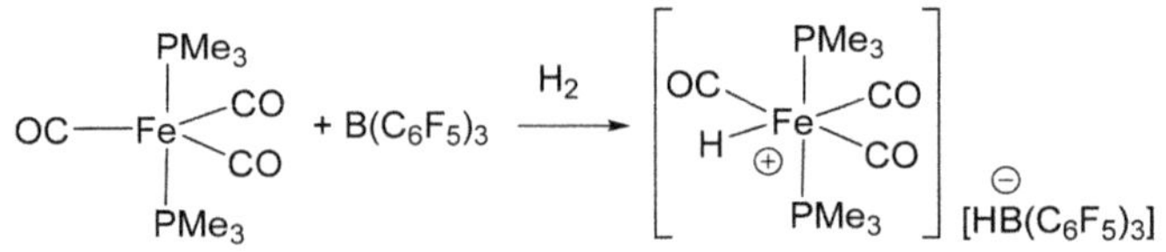

Scheme 5.44 Reaction of Fe(0)/B FLP with dihydrogen.

In a conceptually related example,[77] the iron(0) complex $[Fe(CO)_3(PMe_3)_2]$ was combined with $B(C_6F_5)_3$. The electron-rich metal center and together with the borane act as an FLP in the activation of dihydrogen to give $[HFe(CO)_3(PMe_3)_2][HB(C_6F_5)_3]$ (Scheme 5.44). While the reaction was not reversible, 5 mol% of this FLP under 1 bar of dihydrogen at 130 °C catalyzed the hydrogenation of $Ph_2C=CH_2$ in 76% yield after 24 h. Other terminal olefins were partially reduced (28–48%) while internal ones were not reduced under these mild conditions.

5.15 Bimetallic Systems

In the above sections, we have seen the use of transition metals both as the acidic and basic components of an FLP results in the ability to activate dihydrogen. These observations beg the question of a further step to mixed metal FLPs. In this regard, the work of Mankad and coworkers is relevant. In a 2015 paper, these researchers described bimetallic complexes of the form $(NHC)MM'Cp(CO)_2$ (M = Ag, Cu; M' = Fe, Ru).[78] These compounds catalyzed the hydrogenation of alkynes. Computations regarding the mechanism revealed that the transition state involving dihydrogen showed side-on and end-on interactions of dihydrogen with the Lewis acidic Cu and basic Fe, respectively (Scheme 5.45). This geometry is directly analogous to that observed in the interaction between the encounter complex and dihydrogen P/B FLPs.[79]

In a study of the Mo–Mo species $Mo_2(N(i\text{-}Pr_2C_6H_3)_2CH)_4$ containing a quintuple Mo–Mo bond,[80] it was shown to react with dihydrogen to give the quadruple-bonded dihydride analog. Most interestingly, a computational study revealed that the polarization of the frontier MOs in the transition state resulted in the activation of dihydrogen in a fashion directly analogous to that of an FLP. This finding is certainly interesting as the Mo–Mo quintuple bond is not polar. Moreover, it suggests that an FLP mechanism may be operative in seemingly non-polar systems where a polarized transition state is accessible.

In 2017, Campos[81] described an intermolecular, bimetallic FLP derived from the combination of the bulky phosphine gold Lewis acid $(C_6H_3(2,6\text{-}i\text{-}Pr_2)_2C_6H_3)_2PMe_2AuNTf_2$ and the electron-rich Pt(0) species, $Pt(Pt\text{-}Bu_3)_2$. Exposure of this combination to dihydrogen resulted in an immediate reaction, affording $(C_6H_3(2,6\text{-}i\text{-}Pr_2)_2C_6H_3)_2PMe_2AuH$ and the Pt(0) salt $[HPt(Pt\text{-}Bu_3)_2][N(Tf)_2]$. These initial products slowly interact over 12 h at room temperature to give a hydride-bridged bimetallic species (Scheme 5.46). A subsequent investigation[82] exploiting ligand modification confirmed that steric frustration was required to achieve the activation of dihydrogen, further supporting the description of this system as a metal-only FLP.

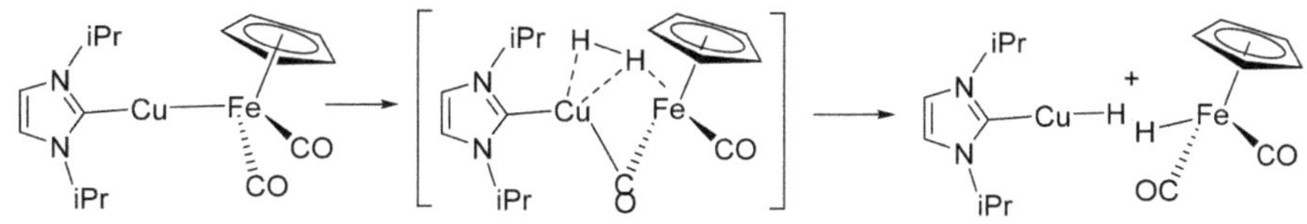

Scheme 5.45 The activation of dihydrogen by a bimetallic species.

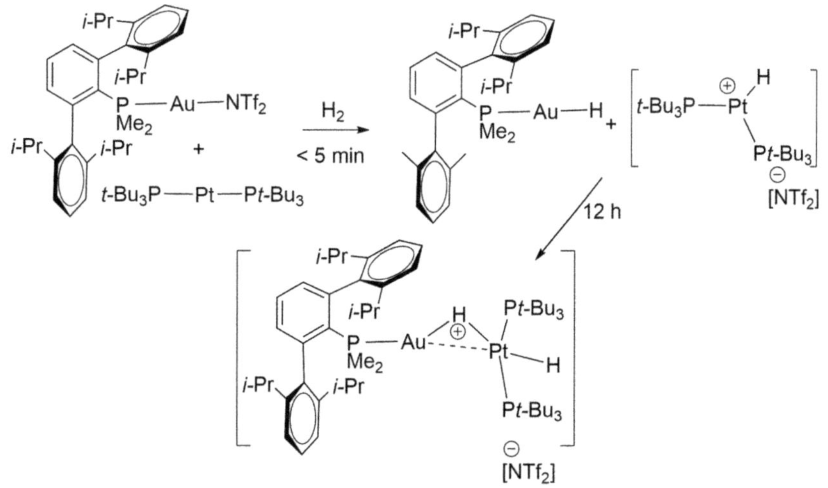

Scheme 5.46 Reaction of an intramolecular all-metal FLP with dihydrogen.

5.16 Alkali Metal Species

Given that transition metals proved viable as components of FLPs in the activation of dihydrogen, our thoughts turned to alkali metal species. Such species are typically comprised of highly basic anions that exhibit weak electrostatic interactions with the Lewis acidic alkali metal center. As such, one can view these species as analogous to FLPs, as there is essentially no covalent bonding between the Lewis acidic and basic sites. This perspective suggested that such species could be capable of dihydrogen activation. Indeed, in the 1960s, the work of Slaugh showed that $LiAlH_4$[83] or suspensions of NaH or KH could be used in the catalytic hydrogenation of olefinic and acetylenic substrates, although high temperatures (150–225 °C) and dihydrogen pressures (60–100 bar) were required.[84] In addition, the subsequent work of Berkessel[85] in 2002 showed catalytic hydrogenation of benzophenone was mediated by KOt-Bu, albeit under harsh conditions (see Section 3.2). This was further affirmed by the work of Harder and coworkers in 2008, when they reported the use of 5 mol% of $C_6H_4(NMe_2)(CHSiMe_3)K$ or 10 mol% of KH in the hydrogenation of diphenylethylene (Scheme 5.47) and phenyl cyclohexene, at 60 °C and 100 bar of dihydrogen pressure. Interestingly the corresponding effort to

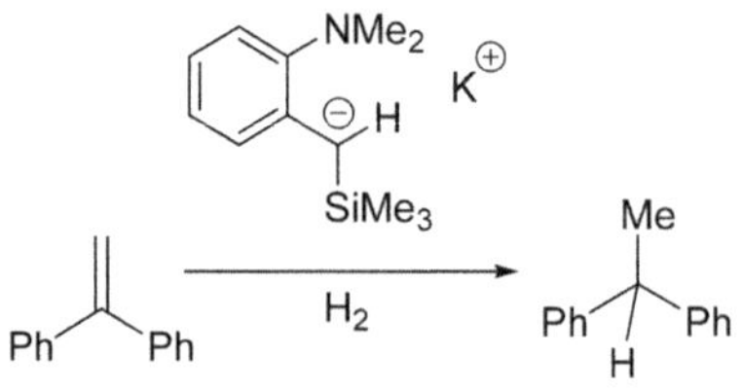

Scheme 5.47 Catalytic hydrogenation of olefin by $C_6H_4(NMe_2)(CHSiMe_3)K$.

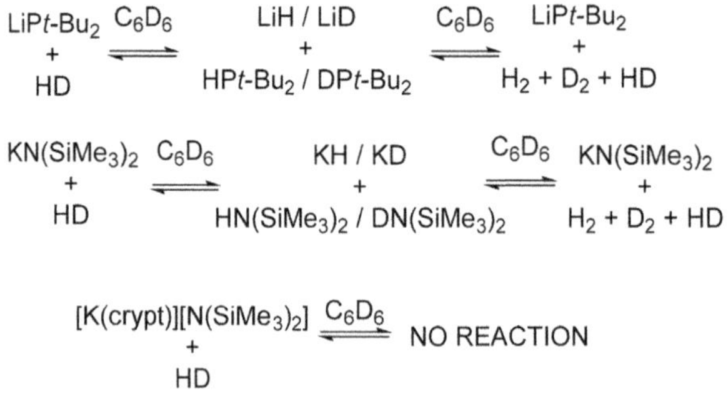

Scheme 5.48 Equilibria involving reaction of $KPt\text{-}Bu_2$ and $K[N(SiMe_3)_2]$ with HD in benzene.

use 5 mol% n-BuLi at 20 °C and 20 bar of dihydrogen was ineffective, affording only 14% of the hydrogenated diphenylethylene.

In a more recent study, we[86] explored the reaction of $LiPt\text{-}Bu_2$ in benzene with HD at 50 °C for 20 h. This led to the formation of an isotopic equilibrium mixture of H_2, D_2, and HD as well as a mixture of $HPt\text{-}Bu_2$ and $DPt\text{-}Bu_2$, as evidenced by NMR spectroscopy. Interestingly, similar mixtures were accessible from combinations of LiH and $HPt\text{-}Bu_2$ under D_2 gas. These data illustrated that the direct reaction of $LiPt\text{-}Bu_2$ and HD effects H–D bond cleavage and, importantly, that the generated LiH(D) and $H(D)Pt\text{-}Bu_2$ reacted to liberate dihydrogen, thus establishing the observed equilibrium mixtures (Scheme 5.48). Similarly, the use of $NaPt\text{-}Bu_2$ or $KPt\text{-}Bu_2$ afforded HD scrambling on heating in toluene after 12 h. The readily available $K[N(SiMe_3)_2]$ behaved similarly, scrambling HD at room temperature in benzene. Even KH was effective at HD scrambling, although refluxing for 40 h in benzene was required. In a control experiment, refluxing KH under N_2 showed no generation of dihydrogen. Perhaps most interestingly, the addition of a cryptand to a solution $K[N(SiMe_3)_2]$ inhibited reaction with HD. This inferred that the Lewis acidity at the alkali metal plays an important role in the activation of the dihydrogen molecule.

Computational studies considered reaction of the dimeric complex $[Li(OEt_2)Pt\text{-}Bu_2]_2$ with dihydrogen proceeds *via* a transition state that involves an interaction of dihydrogen with lithium in a side-on fashion, with concurrent donation from phosphorus to the σ*-orbital of dihydrogen. This geometry of computed transition state TS1 (Figure 5.2) is reminiscent of the activation of dihydrogen by an 'encounter complex' derived from an FLP[79,87–90] and to that described by Mankad for bimetallic systems.[78,91]

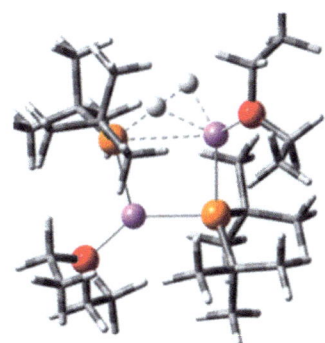

Figure 5.2 Transition state for the reaction of dihydrogen by [LiPt-Bu$_2$]$_2$. O: *red*; P: *orange*, Li: *violet*, H: *grey*. Selected distances: TS1: H···H 1.244 Å, Li···H 1.937, 1.783 Å, P···H 1.634 Å; P···Li 3.059.[86] Reproduced from ref. 86 with permission from John Wiley and Sons, © 2018 Wiley-VCH Verlag GmbH & Co. KGaA, Weinheim.

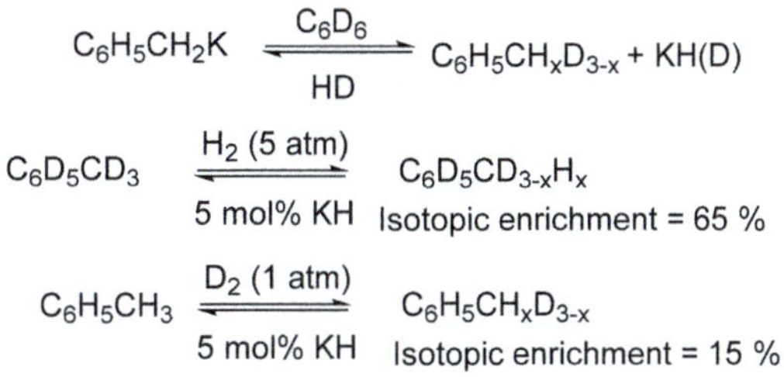

$$C_6H_5CH_2K \xrightleftharpoons[HD]{C_6D_6} C_6H_5CH_xD_{3-x} + KH(D)$$

$$C_6D_5CD_3 \xrightleftharpoons[\text{5 mol\% KH}]{H_2 \text{ (5 atm)}} C_6D_5CD_{3-x}H_x \quad \text{Isotopic enrichment = 65 \%}$$

$$C_6H_5CH_3 \xrightleftharpoons[\text{5 mol\% KH}]{D_2 \text{ (1 atm)}} C_6H_5CH_xD_{3-x} \quad \text{Isotopic enrichment = 15 \%}$$

Scheme 5.49 Reactions of benzyl potassium with HD, isotopic exchange in toluene using KH and dihydrogen and KH and D$_2$.

The reaction of KPt-Bu$_2$ with HD in deuterated toluene as the solvent revealed an additional reaction. The HD scambling as seen in benzene was evident, however the low intensity of the signal from HPt-Bu$_2$ together with the enhanced signal from residual protons in the methyl group of toluene was consistent with H/D exchange among the methyl group of the toluene and HD. This view was affirmed by independent reactions of C$_6$D$_5$CD$_2$K with dihydrogen and C$_6$H$_5$CH$_2$K with D$_2$ which gave isotopic enrichment (Scheme 5.49). These latter reactions were found to be pressure-dependent, and also exhibited a kinetic isotope effect (KIE) consistent with a rate-determining step involving dihydrogen activation.

The ability of the above alkali metal salts to activate dihydrogen prompted the use of these species in hydrogenation catalysis. Indeed, 5 mol% KPPh$_2$ in refluxing THF under dihydrogen (5 bar) catalytically reduced 1,1-diphenylethylene and *N*-benzyl-*tert*-butylimine after 72 and 6 h in 67 and 62% yield, respectively (Scheme 5.50). More recently Pfaltz and coworkers[92] have exploited LiN(SiMe$_3$)$_2$ to affect the hydrogenation of a series of ketimines at 80 °C at 100 bar of dihydrogen.

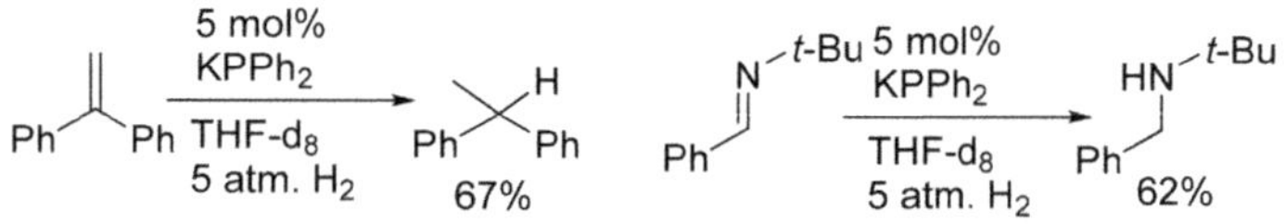

Scheme 5.50 Hydrogenation of substrates with $KPPh_2$ as a catalyst.

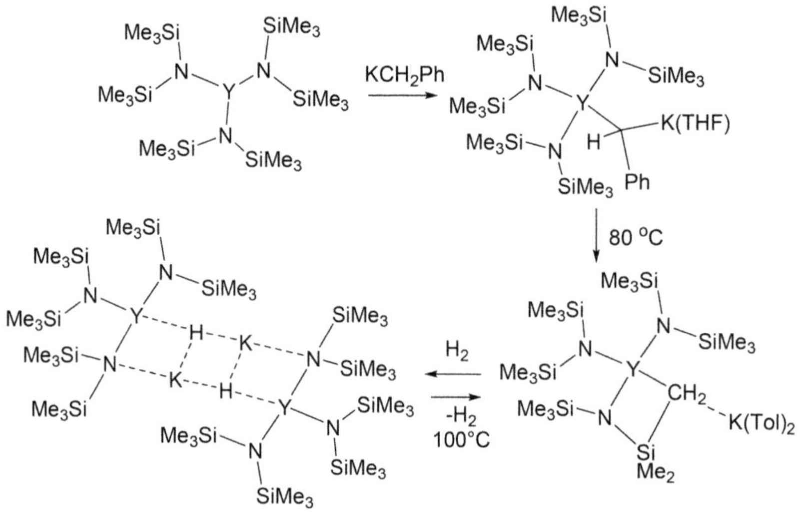

Scheme 5.51 Synthesis and reactions of yttrium benzyl ate complex with dihydrogen.

In a very recent result, the reaction of the yttrium amide complex $((Me_3Si)_2N)_3Y$ with $PhCH_2K$ afforded the benzyl ate complex, $[((Me_3Si)_2N)_3YCHPh(K(THF))]$ which underwent cyclometallation at 80 °C to give $[((Me_3Si)_2N)_2YN(SiMe_3)SiMe_2CH_2K(C_7H_8)_2]$ (Scheme 5.51). The reaction of this latter species with dihydrogen (10 bar) at room temperature generated $[((Me_3Si)_2N)_2Y(\mu-N(SiMe_3)_2)(\mu-H)K]_2$ (Scheme 5.51). This reaction with dihydrogen was reversed by heating to 100 °C. This reversible activation was then exploited to affect the catalytic hydrogenation of alkenes, alkynes, and imines.[93] This activation of dihydrogen is analogous to the activation by $PhCH_2K$ described above as the Y center is ancillary to the proposed activation of dihydrogen between the metalated carbon and the potassium center.

5.17   Alkaline Earth Acids

While MgH_2 had been used to hydrogenate olefinic substrates at high temperatures and pressures as early as the 1960s,[84] it was in 2008 that Harder and co-workers[94] reported an alternative approach to transition metal-free hydrogenation catalysis. These researchers described the use of the alkaline earth complexes $[HC(C(Me)N(2,6-i-Pr_2C_6H_3))_2Ca(THF)H]_2$

and $(C_6H_4(NMe_2)(CHSiMe_3))_2M(THF)_2$ (M = Ca, Sr) to affect the hydrogenation of olefins at temperatures ranging from 20–100 °C and 20–100 bar of dihydrogen pressure (Scheme 5.52).

Subsequently, Okuda's group showed that trimetallic calcium[95] and strontium[96] hydride cations of the form $[((MeN(CH_2)_2)_3(CH_2)_2N)_3M_3(\mu_3\text{-}H)_2]^+$ (M = Ca, Sr) also act as hydrogenation catalysts for the reduction of diphenylethylene at 60 °C and 1 bar of dihydrogen. In the case of the Ca trimetallic species, the reduction is slow, requiring 19 mol% of the catalyst and 13 days to effect complete reduction. In the case of the Sr complex $[((MeN(CH_2)_2)_3(CH_2)_2N)_3Sr_3(\mu_3\text{-}H)_2][SiPh_3]$, this species reacted with dihydrogen in an FLP fashion affording a transient mixture of $((MeN(CH_2)_2)_3(CH_2)_2N)_3Sr_3(\mu_3\text{-}H)_3$ and $HSiPh_3$. The ultimate loss of dihydrogen or benzene gave either the initial Sr complex or the salt $[((MeN(CH_2)_2)_3(CH_2)_2N)_3Sr_3(\mu_3\text{-}H)_2][SiHPh_2]$, respectively (Scheme 5.53). This behavior was confirmed *via* labeling experiments.

More recently Harder *et al.*[97] have also reported the use of $M[N(SiMe_3)_2]_2$, (M = Mg, Ca, Sr, Ba) as hydrogenation catalysts for imine reduction. In this case, the mechanism is thought to involve a catalytic cycle derived from hydrogenolysis of Ca–N bonds, followed by insertion of imine into the Ca–H bond. This is analogous to that described for *i*-Bu$_2$AlH (see Section 5.3). In an interesting and recent advance, the rates of hydrogenation of styrenes and α-olefins were markedly improved employing a mixture catalyst derived from KH and $M[N(SiMe_3)_2]_2$ (M = Ca, Mg) suggesting a synergistic effect.[98]

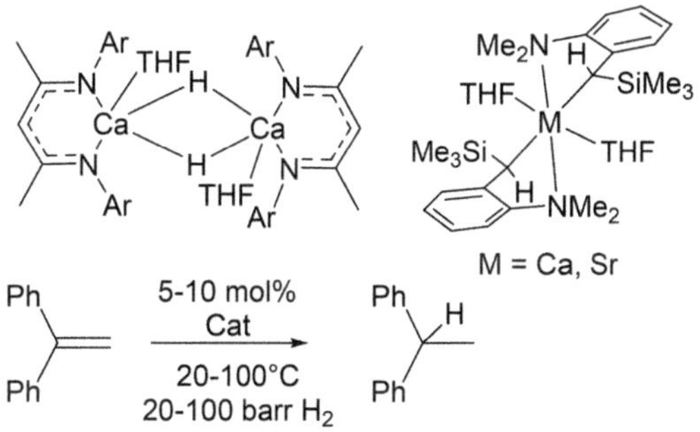

Scheme 5.52 Hydrogenation of olefins by alkaline earth metal complexes.

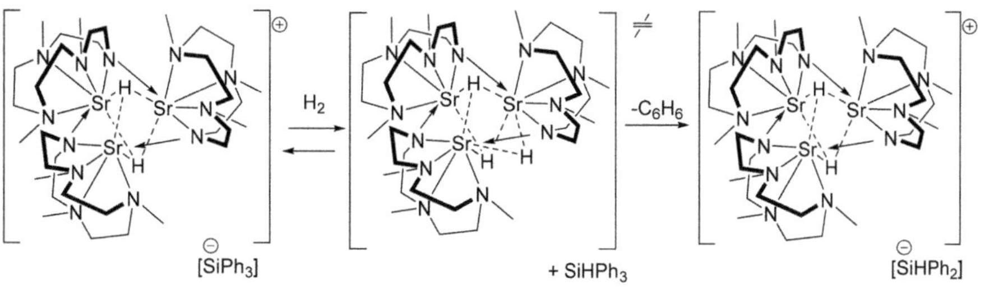

Scheme 5.53 FLP reaction pathway for Sr$_3$ complex with dihydrogen.

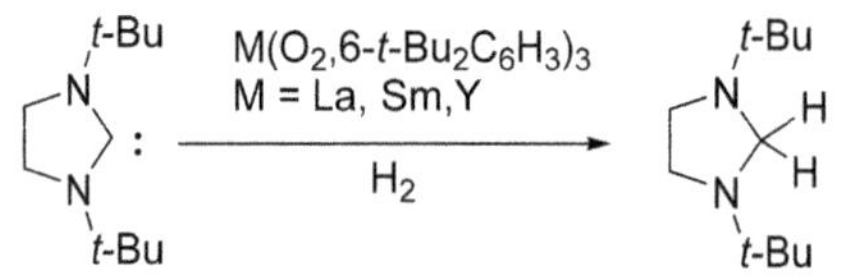

Scheme 5.54 Reaction of carbene with dihydrogen in the presence of rare-earth complexes.

5.18 Rare Earth Acids

The use of rare-earth metals as Lewis acid has also been probed. The reaction of the homoleptic rare-earth metal aryloxides, $M(O(2,6-t-Bu_2C_6H_3))_3$ (M = La, Sm, and Y) in the presence of the carbene $C_3H_4(Nt-Bu)_2$ with dihydrogen under mild conditions gave the aminal $C_3H_6(Nt-Bu)_2$ (Scheme 5.54).[99]

5.19 Implications

This chapter demonstrates that in the 15 years since the inception of FLP chemistry a broad range of systems has been shown to participate in FLP chemistry with dihydrogen. Indeed, a new class of FLPs is derived from the use of Lewis acidic and Lewis basic centers in a general sense. It is also important to recognize the conceptual relation of the FLPs to the well-established Noyori bifunctional catalysts.[100] In the latter systems the activation of dihydrogen takes place across the metal–donor bond. However, in FLPs this concept is generalized to combinations of donors and acceptor molecules. Indeed, the examples in the present chapter build on the original findings, in that they are utilizing new donors or acceptors in the broadest sense. Extensions beyond the main group to transition, alkali, alkaline-earth, and rare earth metals serve to demonstrate the broad applicability of the concept. Perhaps most importantly, this diversity of potential catalysts offers a broad range of possibilities for future development and optimization.

References

1. J. M. Farrell, J. A. Hatnean and D. W. Stephan, *J. Am. Chem. Soc.*, 2012, **134**, 15728–15731.
2. J. M. Farrell, R. T. Posaratnanathan and D. W. Stephan, *Chem. Sci.*, 2015, **6**, 2010–2015.
3. P. Eisenberger, B. P. Bestvater, E. C. Keske and C. M. Crudden, *Angew. Chem., Int. Ed.*, 2015, **54**, 2467–2471.
4. J. M. Farrell and D. W. Stephan, *Chem. Commun.*, 2015, **51**, 14322–14325.
5. E. R. Clark, A. Del Grosso and M. J. Ingleson, *Chem. - Eur. J.*, 2013, **19**, 2462–2466.
6. M. Devillard, R. Brousses, K. Miqueu, G. Bouhadir and D. Bourissou, *Angew. Chem., Int. Ed.*, 2015, **54**, 5722–5726.
7. J. Lam, B. A. R. Gunther, J. M. Farrell, P. Eisenberger, B. P. Bestvater, P. D. Newman, R. L. Melen, C. M. Crudden and D. W. Stephan, *Dalton Trans.*, 2016, **45**, 15303–15316.
8. Y. Segawa, M. Yamashita and K. Nozaki, *Science*, 2006, **314**, 113–115.
9. M. Yamashita and K. Nozaki, *Top. Organomet. Chem.*, 2015, **49**, 1–37.
10. J. H. Zheng, Y. W. Wang, Z. H. Li and H. D. Wang, *Chem. Commun.*, 2015, **51**, 5505–5508.
11. L. Liu, C. Chan, J. Zhu, C. H. Cheng and Y. Zhao, *J. Org. Chem.*, 2015, **80**, 8790–8795.
12. D. Wu, L. B. Kong, Y. X. Li, R. Ganguly and R. Kinjo, *Nat. Commun.*, 2015, **6**, 7340.

13. M. A. Dureen and D. W. Stephan, *J. Am. Chem. Soc.*, 2009, **131**, 8396–8397.
14. G. Ménard and D. W. Stephan, *Angew. Chem., Int. Ed.*, 2012, **51**, 8272–8275.
15. G. Ménard, L. Tran and D. W. Stephan, *Dalton Trans.*, 2013, **42**, 13685–13691.
16. M. Xu, J. Possart, A. E. Waked, J. Roy, W. Uhl and D. W. Stephan, *Philos. Trans. R. Soc., A*, 2017, **375**, 2101.
17. J. A. Hatnean, J. W. Thomson, P. A. Chase and D. W. Stephan, *Chem. Commun.*, 2014, **50**, 301–303.
18. G. Schnee, O. Nieto Faza, D. Specklin, B. Jacques, L. Karmazin, R. Welter, C. Silva López and S. Dagorne, *Chem. - Eur. J.*, 2015, **21**, 17959–17972.
19. A. Bolley, G. Schnee, L. Thevenin, B. Jacques and S. Dagorne, *Inorganics*, 2018, **6**, 23/21–23/29.
20. A. Bodach, N. Noethling and M. Felderhoff, *Eur. J. Inorg. Chem.*, 2021, **2021**(13), 1240–1243, Ahead of Print.
21. D. W. N. Wilson, J. Feld and J. M. Goicoechea, *Angew. Chem., Int. Ed.*, 2020, **59**, 20914–20918.
22. S. Harder, A. Friedrich, J. Eyselein, H. Elsen, J. Langer, J. Pahl and M. Weisinger, *Chem. - Eur. J.*, 2021, **27**(28), 7756–7763, accepted.
23. G. D. Frey, V. Lavallo, B. Donnadieu, W. W. Schoeller and G. Bertrand, *Science*, 2007, **316**, 439–441.
24. J. W. Runyon, O. Steinhof, H. V. R. Dias, J. C. Calabrese, W. J. Marshall and A. J. Arduengo, *Aust. J. Chem.*, 2011, **64**, 1165–1172.
25. E. R. Clark and M. J. Ingleson, *Organometallics*, 2013, **32**, 6712–6717.
26. E. R. Clark and M. J. Ingleson, *Angew. Chem., Int. Ed.*, 2014, **53**, 11306–11309.
27. M. P. Boone and D. W. Stephan, *J. Am. Chem. Soc.*, 2013, **135**, 8508–8511.
28. C. Fan, L. G. Mercier, W. E. Piers, H. M. Tuononen and M. Parvez, *J. Am. Chem. Soc.*, 2010, **132**, 9604–9606.
29. M. Alcarazo, C. Gomez, S. Holle and R. Goddard, *Angew. Chem., Int. Ed.*, 2010, **49**, 5788–5791.
30. J. Cabrera-Trujillo Jorge and I. Fernandez, *Inorg. Chem.*, 2019, **58**, 7828–7836.
31. H. Li, A. J. A. Aquino, D. B. Cordes, F. Hung-Low, W. L. Hase and C. Krempner, *J. Am. Chem. Soc.*, 2013, **135**, 16066–16069.
32. P. Cui, C. C. Comanescu and V. M. Iluc, *Chem. Commun.*, 2015, **51**, 6206–6209.
33. A. Schafer, M. Reissmann, A. Schafer, W. Saak, D. Haase and T. Muller, *Angew. Chem., Int. Ed.*, 2011, **50**, 12636–12638.
34. M. Reissmann, A. Schafer, S. Jung and T. Müller, *Organometallics*, 2013, **32**, 6736–6744.
35. T. J. Herrington, B. J. Ward, L. R. Doyle, J. McDermott, A. J. P. White, P. A. Hunt and A. E. Ashley, *Chem. Commun.*, 2014, **50**, 12753.
36. A. Schäfer, M. Reißmann, A. Schäfer, M. Schmidtmann and T. Müller, *Chem. - Eur. J.*, 2014, **20**, 9381.
37. Z. Dong, Z. Li, X. Liu, C. Yan, N. Wei, M. Kira, Z. Dong and T. Muller, *Chem.–Asian J.*, 2017, **12**, 1204–1207.
38. Y. Wang, M. Karni, S. Yao, A. Kaushansky, Y. Apeloig and M. Driess, *J. Am. Chem. Soc.*, 2019, **141**, 12916–12927.
39. B. J. Guddorf, A. Hepp, C. Daniliuc, D. W. Stephan and F. Lips, *Dalton Trans.*, 2020, **49**, 13386–13392.
40. R. T. Cooper, J. S. Sapsford, R. C. Turnell-Ritson, D. H. Hyon, A. J. P. White and A. E. Ashley, *Philos. Trans. R. Soc., A*, 2017, **375**, 2101.
41. D. J. Scott, N. A. Phillips, J. S. Sapsford, A. C. Deacy, M. J. Fuchter and A. E. Ashley, *Angew. Chem., Int. Ed.*, 2016, **55**, 14738–14742.
42. S. Das, S. Mondal and S. K. Pati, *Chem. - Eur. J.*, 2018, **24**, 2575–2579.
43. S. Das and S. K. Pati, *Catal. Sci. Technol.*, 2018, **8**, 5178–5189.
44. S. Das and S. K. Pati, *Organometallics*, 2021, **40**, 194–202.
45. J. S. Sapsford, D. Csokas, D. J. Scott, R. C. Turnell-Ritson, A. D. Piascik, I. Papai and A. E. Ashley, *ACS Catal.*, 2020, **10**, 7573–7583.
46. P. Holtkamp, J. Schwabedissen, B. Neumann, H.-G. Stammler, I. V. Koptyug, V. V. Zhivonitko and N. W. Mitzel, *Chem. - Eur. J.*, 2020, **26**, 17381–17385.
47. L. E. Longobardi, C. A. Russell, M. Green, N. S. Townsend, K. Wang, A. J. Holmes, S. B. Duckett, J. E. McGrady and D. W. Stephan, *J. Am. Chem. Soc.*, 2014, **136**, 13453–13457.
48. S. S. Chitnis, A. P. M. Robertson, N. Burford, B. O. Patrick, R. McDonald and M. J. Ferguson, *Chem. Sci.*, 2015, **6**, 6545–6555.
49. L. J. Hounjet, C. B. Caputo and D. W. Stephan, *Dalton Trans.*, 2013, **42**, 2629–2635.
50. C. B. Caputo, L. J. Hounjet, R. Dobrovetsky and D. W. Stephan, *Science*, 2013, **341**, 1374–1377.
51. T. vom Stein, M. Pérez, R. Dobrovetsky, D. Winkelhaus, C. B. Caputo and D. W. Stephan, *Angew. Chem., Int. Ed.*, 2015, **54**, 10178–10182.
52. L. Cabrera, E. Hollink, J. C. Stewart, P. R. Wei and D. W. Stephan, *Organometallics*, 2005, **24**, 1091–1098.
53. A. M. Chapman, M. F. Haddow and D. F. Wass, *J. Am. Chem. Soc.*, 2011, **133**, 8826–8829.
54. A. M. Chapman, M. F. Haddow and D. F. Wass, *J. Am. Chem. Soc.*, 2011, **133**, 18463–18478.

55. A. M. Chapman and D. F. Wass, *Dalton Trans.*, 2012, **41**, 9067–9072.
56. A. M. Chapman, M. F. Haddow and D. F. Wass, *Eur. J. Inorg. Chem.*, 2012, **2012**, 1546–1554.
57. S. R. Flynn, O. J. Metters, I. Manners and D. F. Wass, *Organometallics*, 2016, **35**, 847–850.
58. X. Xu, G. Kehr, C. G. Daniliuc and G. Erker, *J. Am. Chem. Soc.*, 2015, **137**, 4550–4557.
59. X. Xu, G. Kehr, C. G. Daniliuc and G. Erker, *J. Am. Chem. Soc.*, 2013, **135**, 6465–6476.
60. Z. B. Jian, C. G. Daniliuc, G. Kehr and G. Erker, *Organometallics*, 2017, **36**, 424–434.
61. A. T. Normand, C. G. Daniliuc, B. Wibbeling, G. Kehr, P. Le Gendre and G. Erker, *J. Am. Chem. Soc.*, 2015, **137**, 10796–10808.
62. P. H. M. Budzelaar, D. L. Hughes, M. Bochmann, A. Macchioni and L. Rocchigiani, *Chem. Commun.*, 2020, **56**, 2542–2545.
63. C. J. Curtis, A. Miedaner, W. W. Ellis and D. L. Dubois, *J. Am. Chem. Soc.*, 2002, **124**, 1918–1925.
64. S. Raugei, S. T. Chen, M. H. Ho, B. Ginovska-Pangovska, R. J. Rousseau, M. Dupuis, D. L. DuBois and R. M. Bullock, *Chem. - Eur. J.*, 2012, **18**, 6493–6506.
65. R. M. Bullock and M. L. Helm, *Acc. Chem. Res.*, 2015, **48**, 2017–2026.
66. S. G. Zhang, A. M. Appel and R. M. Bullock, *J. Am. Chem. Soc.*, 2017, **139**, 7376–7387.
67. T.-T. Thai, D. S. Mérel, A. Poater, S. Gaillard and J.-L. Renaud, *Chem. - Eur. J.*, 2015, **21**, 7066–7070.
68. A. Lator, Q. G. Gaillard, D. S. Merel, J.-F. Lohier, S. Gaillard, A. Poater and J.-L. Renaud, *J. Org. Chem.*, 2019, **84**, 6813–6829.
69. Y. Lin, D.-P. Zhu, Y.-R. Du, R. Zhang, S.-J. Zhang and B.-H. Xu, *Org. Lett.*, 2019, **21**, 2693–2698.
70. C. Cristobal, Y. A. Hernandez, J. Lopez-Serrano, M. Paneque, A. Petronilho, M. L. Poveda, V. Salazar, F. Vattier, E. Alvarez, C. Maya and E. Carmona, *Chem. - Eur. J.*, 2013, **19**, 4003–4020.
71. M. Carmona, J. Ferrer, R. Rodriguez, V. Passarelli, F. J. Lahoz, P. Garcia-Orduna, L. Canadillas-Delgado and D. Carmona, *Chem. - Eur. J.*, 2019, **25**, 13665–13670.
72. S. K. Podiyanachari, R. Froehlich, C. G. Daniliuc, J. L. Petersen, C. Mueck-Lichtenfeld, G. Kehr and G. Erker, *Angew. Chem., Int. Ed.*, 2012, **51**, 8830–8833.
73. W. H. Harman and J. C. Peters, *J. Am. Chem. Soc.*, 2012, **134**, 5080–5082.
74. S. J. K. Forrest, J. Clifton, N. Fey, P. G. Pringle, H. A. Sparkes and D. F. Wass, *Angew. Chem., Int. Ed.*, 2015, **54**, 2223–2227.
75. K. Mistry, P. G. Pringle, H. A. Sparkes and D. F. Wass, *Organometallics*, 2020, **39**, 468–477.
76. B. R. Barnett, C. E. Moore, A. L. Rheingold and J. S. Figueroa, *J. Am. Chem. Soc.*, 2014, **136**, 10262–10265.
77. H. Tinnermann, C. Fraser and R. D. Young, *Dalton Trans.*, 2020, **49**, 15184–15189.
78. M. K. Karunananda and N. P. Mankad, *J. Am. Chem. Soc.*, 2015, **137**, 14598–14601.
79. S. Grimme, H. Kruse, L. Goerigk and G. Erker, *Angew. Chem., Int. Ed.*, 2010, **49**, 1402–1405.
80. Y. Chen and S. Sakaki, *Inorg. Chem.*, 2017, **56**, 4011–4020.
81. J. Campos, *J. Am. Chem. Soc.*, 2017, **139**, 2944–2947.
82. N. Hidalgo, J. J. Moreno, M. Pérez-Jiménez, C. Maya, J. López-Serrano and J. Campos, *Chem. - Eur. J.*, 2020, **26**, 5982–5993.
83. L. H. Slaugh, *Tetrahedron*, 1966, **22**, 1741–1746.
84. L. H. Slaugh, *J. Org. Chem.*, 1967, **32**, 108–113.
85. A. Berkessel, T. J. S. Schubert and T. N. Mueller, *J. Am. Chem. Soc.*, 2002, **124**, 8693–8698.
86. M. Xu, A. R. Jupp, Z. W. Qu and D. W. Stephan, *Angew. Chem., Int. Ed.*, 2018, **57**, 11050–11054.
87. G. C. Welch, R. R. S. Juan, J. D. Masuda and D. W. Stephan, *Science*, 2006, **314**, 1124–1126.
88. G. C. Welch and D. W. Stephan, *J. Am. Chem. Soc.*, 2007, **129**, 1880–1881.
89. T. A. Rokob, A. Hamza, A. Stirling, T. Soós and I. Papai, *Angew. Chem., Int. Ed.*, 2008, **47**, 2435–2438.
90. T. A. Rokob, A. Hamza, A. Stirling and I. Papai, *J. Am. Chem. Soc.*, 2009, **131**, 2029–2036.
91. N. P. Mankad, *Chem. - Eur. J.*, 2016, **22**, 5822–5829.
92. D. C. Elliott, A. Marti, P. Mauleón and A. Pfaltz, *Chem. - Eur. J.*, 2019, **25**, 1918–1922.
93. D.-D. Zhai, H.-Z. Du, X.-Y. Zhang, Y.-F. Liu and B.-T. Guan, *ACS Catal.*, 2019, **9**, 8766–8771.
94. J. Spielmann, F. Buch and S. Harder, *Angew. Chem., Int. Ed.*, 2008, **47**, 9434–9438.
95. P. Jochmann, J. P. Davin, T. P. Spaniol, L. Maron and J. Okuda, *Angew. Chem., Int. Ed.*, 2012, **51**, 4452–4455.
96. D. Mukherjee, T. Hollerhage, V. Leich, T. P. Spaniol, U. Englert, L. Maron and J. Okuda, *J. Am. Chem. Soc.*, 2018, **140**, 3403–3411.
97. H. Bauer, M. Alonso, C. Färber, H. Elsen, J. Pahl, A. Causero, G. Ballmann, F. d. Proft and S. Harder, *Nat. Catal.*, 2018, **1**, 40–47.
98. X.-Y. Zhang, H.-Z. Du, D.-D. Zhai and B.-T. Guan, *Org. Chem. Front.*, 2020, **7**, 1991–1996.
99. K. Chang, Y. Dong and X. Xu, *Chem. Commun.*, 2019, **55**, 12777–12780.
100. R. Noyori, *Angew. Chem., Int. Ed.*, 2002, **41**, 2008–2022.

6 Other Directions for FLP Hydrogenations

6.1 Chapter Overview

The previous five chapters have described the evolution of FLP chemistry from discovery to applications in the hydrogenation of a range of organic substrates. The emergence of the further sophistication of asymmetric reductions as well as the broadening of FLP systems to cover combinations of Lewis acids and bases derived from compounds across the periodic table have also been described. These findings establish the generality of the concept of FLPs as well as provide access to a diverse array of new hydrogenation catalysts.

With this background, we are positioned to exploit the concept of FLPs in new and unexplored ways. Herein, we describe efforts that apply the principle in directions well beyond organic reductions. Some of the aspects considered offer logical expansions of the reactivity to include the participation or generation of radicals, transfer hydrogenations, dehydrogenation, and the applications of hydrogenation to small-molecule substrates. In addition, we also discuss other areas of growth where the relation to FLP chemistry is perhaps less obvious. The relevance to enzymatic systems, efforts to develop supported FLP catalysts, solid-state reactivity, and pertinence to heterogeneous catalysts are also considered. These additional directions of growth and applications of FLP–dihydrogen reactivity further affirm the potential of FLP chemistry as a new axiom of chemical reactivity.

6.2 Radicals in FLP Reactions with Dihydrogen

The concept of FLPs so far has been based on the combination of two-electron donors and two-electron acceptors. One direction for conceptual expansion involves the notion of utilizing one-electron donors or acceptors. While this possibility was previously inferred by the consideration of a radical mechanism of action (see Section 2.6),

A Primer in Frustrated Lewis Pair Hydrogenation: Concepts to Applications
By Douglas W. Stephan
© Douglas W. Stephan 2022
Published by the Royal Society of Chemistry, www.rsc.org

the earlier results did not provide direct evidence for the participation of radicals in reactions with dihydrogen. However, in 2016, Erker and coworkers[1] probed the reaction of the persistent radical TEMPO, $C_5H_6Me_4NO^{\cdot}$ in the presence of $B(C_6F_5)_3$ and dihydrogen. This combination generated $C_5H_6Me_4NOH$, which underwent further reaction with the borane to ultimately afford the salt $[C_4H_6Me_2N=CMe_2][(HO)(B(C_6F_5)_3)_2]$ in addition to the zwitterion $C_5H_6Me_4N(H)OB(C_6F_5)_3$ (Scheme 6.1). This reaction was thought to proceed by a typical FLP mechanism in which TEMPO acted as the Lewis base. Subsequent ring contraction or proton migration afforded the two observed products. DFT data revealed there is no significant spin density on the hydrogen atoms consistent with a mechanism involving heterolytic cleavage.

Another avenue to radical chemistry was uncovered in our work,[2] targeting the FLP hydrogenation of 9,10-phenanthrenequinone. In this case, the dione was reacted with 1 equivalent of $B(C_6F_5)_3$ at room temperature under dihydrogen, affording two products. The major product was the green–black, air-stable radical $[[(C_6F_5)_2B(O_2C_{14}H_8)]^{\cdot}$, while the minor product was the boronic ester $(C_6F_5)B(O_2C_{14}H_8)$. The analogous products were also derived from 4,5-pyrenedione. DFT calculations revealed the mechanism involved FLP hydrogenation of the dione. The transient diol reacted with the precursor dione to generate the known 9,10-phenanthrene quinhydrone radical.[3] Coordination to $B(C_6F_5)_3$ leads to loss of C_6F_5H, providing the major product radical. The direct reaction of the transient diol with $B(C_6F_5)_3$ prompted protonolysis affording the minor, boronic ester product (Scheme 6.2).

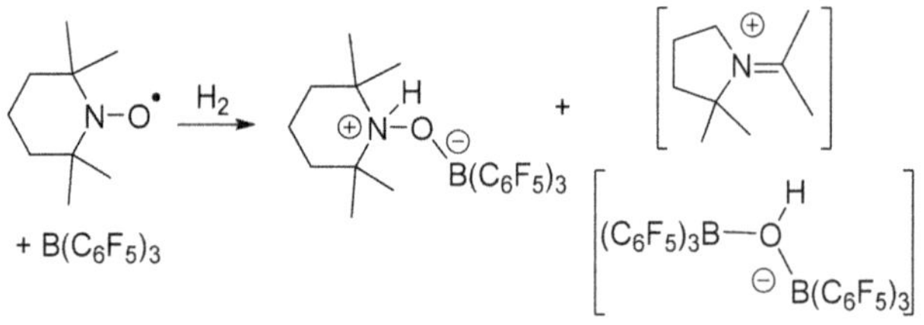

Scheme 6.1   Reactions of TEMPO, $B(C_6F_5)_3$ and dihydrogen.

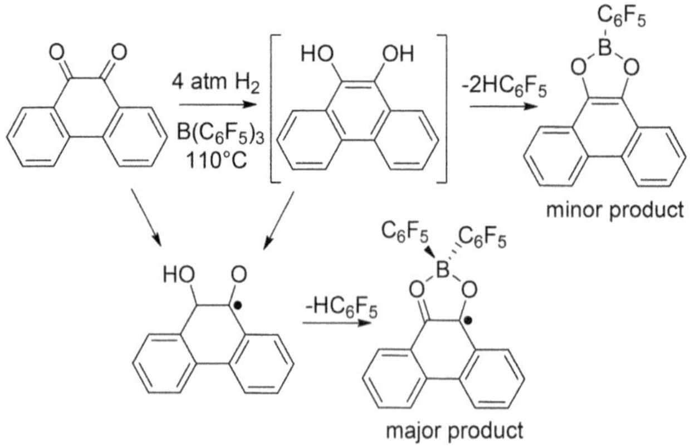

Scheme 6.2   Reaction of 9,10-phenanthrenequinone with $B(C_6F_5)_3$ and dihydrogen.

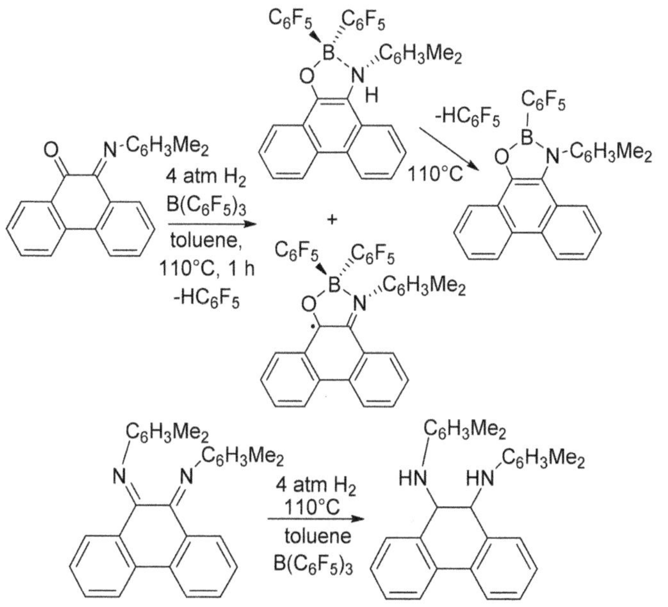

Scheme 6.3 Reactions of iminoquinones and diimines with $B(C_6F_5)_3$ and dihydrogen.

In related reactions[4] of phenanthrene *ortho*-iminoquinones under dihydrogen in the presence of $B(C_6F_5)_3$, the analogous radical product, $[(C_6F_5)_2B((2,6\text{-}Me_2C_6H_3N) OC_{14}H_8)]^{\bullet}$, was formed in addition to the diamagnetic species $[(C_6F_5)_2B((2,6\text{-}Me_2 C_6H_3NH)OC_{14}H_8)]$ (Scheme 6.3). Prolonged heating of the latter species resulted in the loss of C_6F_5H and the formation of $[C_6F_5B((2,6\text{-}Me_2C_6H_3N)OC_{14}H_8)]$. The corresponding diimine derivative was hydrogenated on exposure to dihydrogen in the presence of $B(C_6F_5)_3$. The diamine product was formed with no evidence of a radical intermediate.

6.3 Transfer Hydrogenations

While catalytic hydrogenations are atom-economic with the use of molecular dihydrogen, there are always safety concerns regarding the issues of gas under pressure and flammability. Thus, the use of small molecule sources of dihydrogen offers some safety and practical advantages. This latter avenue is known as 'transfer hydrogenation' and has been widely developed in transition metal-mediated hydrogenations, often using isopropanol or diisopropylamine as the source of dihydrogen.[5–7] As described earlier (see Section 4.2), $B(C_6F_5)_3$ can abstract hydride from the carbon atoms *alpha* to nitrogen.[8–10] This observation suggested that *i*-Pr$_2$NH could act as a source of dihydrogen. Indeed, the transfer hydrogenation of imines, enamines, quinolines, and aziridines mediated by 1–20 mol% $B(C_6F_5)_3$ in the presence of a 100-fold excess of *i*-Pr$_2$NH resulted in reductions that were near quantitative after 24 h at 100 °C (Scheme 6.4).

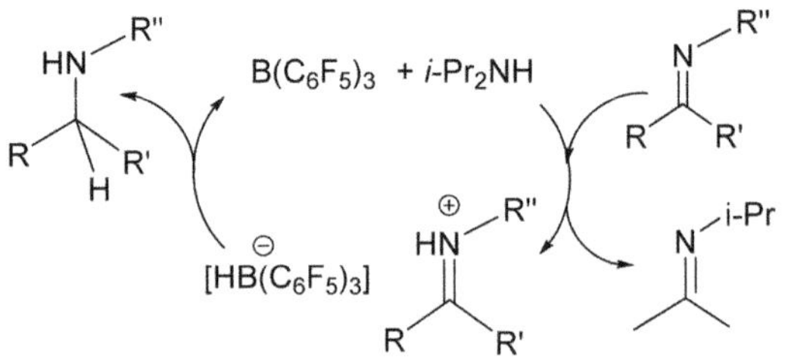

Scheme 6.4   B(C$_6$F$_5$)$_3$ catalyzed transfer hydrogenation.

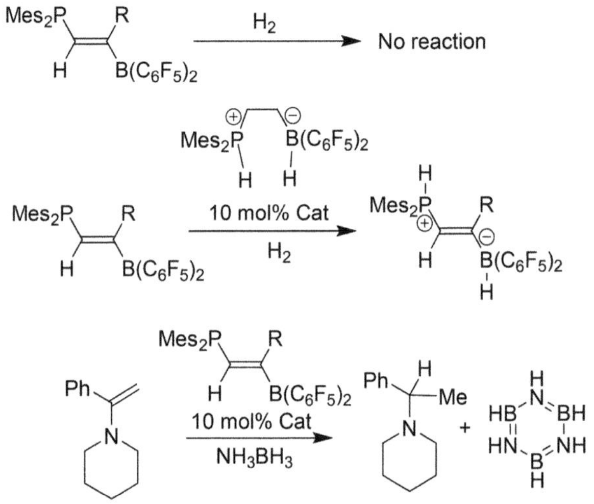

Scheme 6.5   Transfer hydrogenations involving Mes$_2$PCH=CRB(C$_6$F$_5$)$_2$.

Another example of transfer hydrogenation was reported by the Erker group. They had developed the synthesis of the intramolecular FLPs Mes$_2$PCH=CRB(C$_6$F$_5$)$_2$ (R = Me, Ph) and showed that these species did not react with dihydrogen, even at elevated pressures of 60 bar. However, treatment with 10 mol% of the zwitterion, Mes$_2$PHCH$_2$CH$_2$BH(C$_6$F$_5$)$_2$ under dihydrogen (2.5 bar) affected the transfer hydrogenation of Mes$_2$PCH=CRB(C$_6$F$_5$)$_2$, affording Mes$_2$PHCH=CRBH(C$_6$F$_5$)$_2$.[11] As this product was not accessible from reaction with dihydrogen, the species Mes$_2$PCH=CRB(C$_6$F$_5$)$_2$ was precluded from use in conventional catalytic hydrogenations. However, the zwitterion was accessible using ammonia-borane (H$_3$NBH$_3$) as the source of dihydrogen. Thus, the species Mes$_2$PCH=CMeB(C$_6$F$_5$)$_2$ (10–16 mol%) could be employed in the transfer hydrogenation of the enamine PhNC$_5$H$_{10}$C=CH$_2$ affording the expected tertiary amine and the borazine byproduct (Scheme 6.5).

Efforts to apply FLP dehydrogenation in organic applications has also drawn some attention. A computational study probed the potential of B/N intramolecular FLPs in the dehydrogenation of alcohols.[12] On the other hand, Paradies and coworkers[13] have described the dehydrogenation of N-protected indolines, as well as 1,4-dihydropyridines, thiazoline, isoindoline, or 1,2-dihydroquinolines, employing 5 mol% of $B(C_6F_5)_3$ to affect the loss of dihydrogen (Scheme 6.6). Mechanistic studies showed that the rate-determining step was the liberation of dihydrogen from the intermediate ammonium hydridoborate. The addition of the weaker Lewis acid $B(2,4,6\text{-}F_3C_6H_2)_3$ accelerated the reaction by acting as a hydride shuttle and facilitating loss of dihydrogen.

In a more recent and more sophisticated application to organic synthesis, Maji and coworkers[14] have described the dehydrogenative cyclization of *N*-tosyl hydrazones with aromatic amines mediated by $B(C_6F_5)_3$ (Scheme 6.7). In this fashion these authors have catalytically prepared a series of 30 3,4,5-triaryl-1,2,4-triazoles. Experimental and computational studies suggest that Lewis adducts of *N*-tosyl hydrazone with $B(C_6F_5)_3$ begin with intermolecular amination with aniline, which amounts to FLP addition to the hydrazone C=N bond. This is followed by intramolecular cyclization and FLP mediated dehydrogenation to yield the substituted 1,2,4-triazoles (Scheme 6.7).

Transfer hydrogenation has also been extended to carbon-based cationic Lewis acids. Clark and Ingleson[15] showed that such imine reduction was significantly improved using Me_2NHBH_3 as a source of dihydrogen. Thus, the imine $PhC(H)=Nt\text{-}Bu$ was reduced quantitatively in the presence of 5 mol% of *N*-methyl acridinium salt,

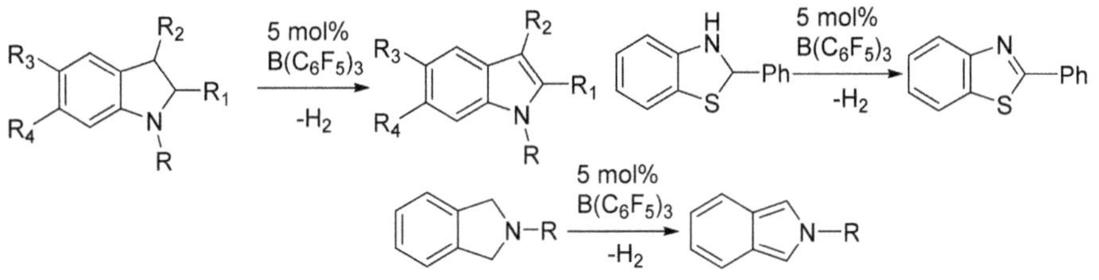

Scheme 6.6 FLP-dehydrogenation of 1,4-dihydropyridines, thiazoline, and isoindoline.

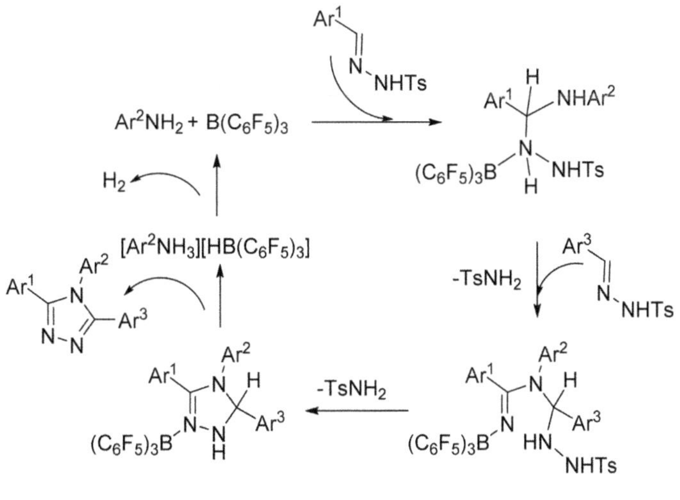

Scheme 6.7 Dehydrogenative cyclization of *N*-tosyl hydrazones.

$[C_{13}H_9NMe][B(3,5-Cl_2C_6H_3)_4]$ and Me_2NHBH_3 at 60 °C in 18 h (Scheme 6.8). This contrasted with the direct hydrogenation using dihydrogen, where this cation showed poor reactivity (See Section 5.5).

Other FLP-derived transfer hydrogenations were derived using electrophilic phosphonium cations as the Lewis acids.[16] Such species were shown to mediate the hydrosilylation of alkenes and alkynes,[17] and dehydrocoupling of amines, thiols, phenols and carboxylic acids with silanes.[18] However, in the latter case, the performance of the reactions in the presence of olefins affected both the dehydrocoupling catalysis as well as the catalytic reduction of the olefin. The net effect was the transfer of hydride and proton from silane and the protic reagent (amine, thiol, alcohol, acid) to the olefin with the simultaneous formation of a Si–E (E = N, S, O) bond (Scheme 6.9).

Oestreich and Chatterjee[19] have exploited cyclohexadienes as effective sources of dihydrogen for the $B(C_6F_5)_3$ mediated transfer hydrogenation of imines and acridines. More recently, the Oestreich group[20] extended the use of cyclohexadienes to the $B(C_6F_5)_3$ catalyzed reduction of 1,1-diaryl olefins. In a subsequent paper,[21] the team of Merrill and Melen optimized such borane mediated transfer hydrogenations for silyl enol ethers, using γ-terpinene in the presence of $C_5H_6Me_4NH$ and $B(C_6F_5)_3$. Hydrolysis of the reduced products afforded access to 29 alcohols. In mechanistic studies, Oestreich *et al.*[20] revealed the process was initiated by hydride abstraction from the cyclohexadiene, affording $[HB(C_6F_5)_3]^-$ and the Brønsted acidic, Wheland cation. This prompted protonation of the alkene and subsequent hydride delivery to the carbocation affording the alkane product and regeneration of the borane catalyst (Scheme 6.10).

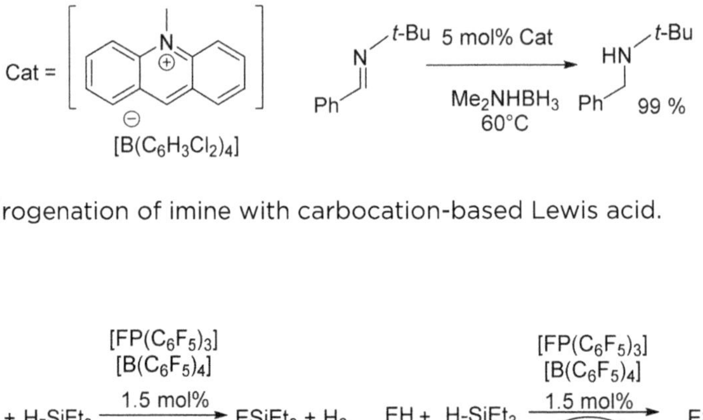

Scheme 6.8 Hydrogenation of imine with carbocation-based Lewis acid.

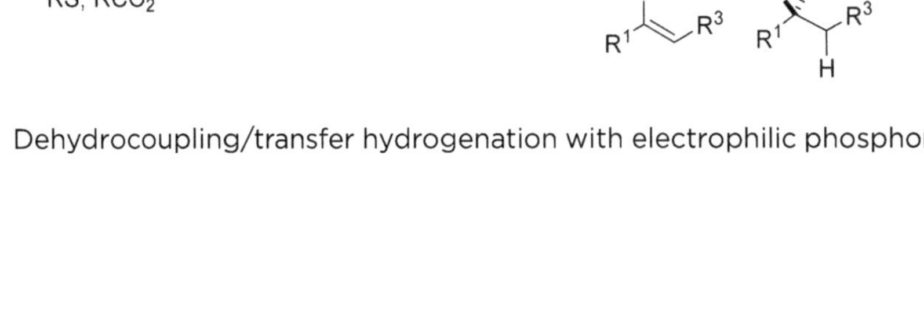

Scheme 6.9 Dehydrocoupling/transfer hydrogenation with electrophilic phosphonium cation.

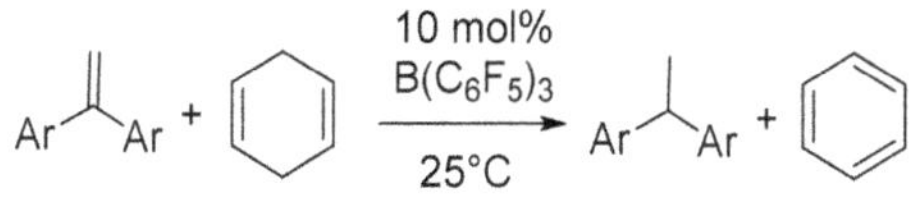

Scheme 6.10   Transfer hydrogenation using cyclohexadiene.

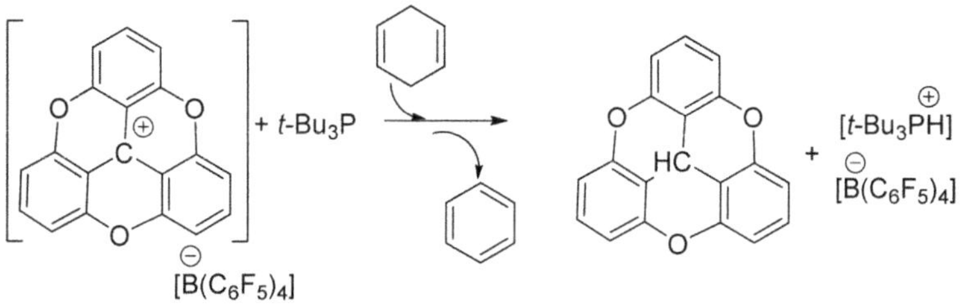

Scheme 6.11   Reaction of $[(C_6H_3O)_3C][B(C_6F_5)_4]$ with phosphine and dihydrogen.

   Gianetti and coworkers[22] developed a novel Lewis acid based on the constrained trityl salt $[(C_6H_3O)_3C][B(C_6F_5)_4]$. This species formed FLPs with bulky phosphines which affected the dehydrogenation of cyclohexadiene to give benzene, the triphenyl-methane, $[(C_6H_3O)_3CH$ and the salt $[t\text{-}Bu_3PH][B(C_6F_5)_4]$ (Scheme 6.11).

6.4   Asymmetric Transfer Hydrogenations

The establishment of transfer hydrogenation as an alternative approach to organic reductions prompted the Du group[23] to target the use of an alternate source of hydrogen for asymmetric reductions. These authors employed a combination of Piers' borane, $HB(C_6F_5)_2$, and (*S*)-*tert*-butylsulfinamide to generate a chiral catalyst *in situ*. Efforts to use these catalysts to mediate the hydrogenate of the imine, Ph(Me)C=NPh using dihydrogen as the reductant resulted in only 10% yield of the amine, although the enantiomeric excess was 85%. However, in sharp contrast, this catalyst mediated the hydrogenation in 95% yield when ammonia-borane was used as the source of dihydrogen, although the enantiomeric excess was reduced to 35%. Nonetheless, applying this latter transfer hydrogenation protocol to other ketimines proved effective, affording an avenue to optically active amines in yields of 78–99% with enantiomeric excesses ranging from 84–95% (Scheme 6.12). The observed reaction was proposed to result from the interaction of the sulfoxide–borane adduct with either ammonia-borane or imine to affect the transfer hydrogenations. Calculations suggested chelation led to a preferred diastereomeric intermediate, affording the preferred enantiomer of the product amine.

   Similarly, employing $HB(C_6F_5)_2$ and (*R*)-*tert*-butylsulfinamide and ammonia borane as the dihydrogen source, *cis*-tetrahydro quinoxalines were obtained in 77–86% enantiomeric excess from the transfer hydrogenation of 2-alkyl-3-arylquinoxaline substrates

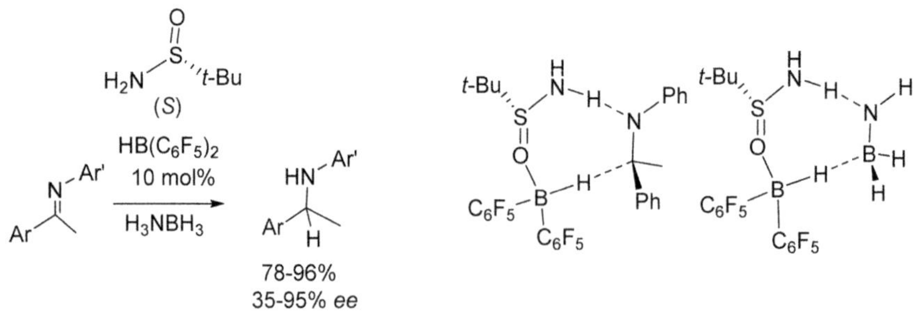

Scheme 6.12 Asymmetric transfer hydrogenation of imines.

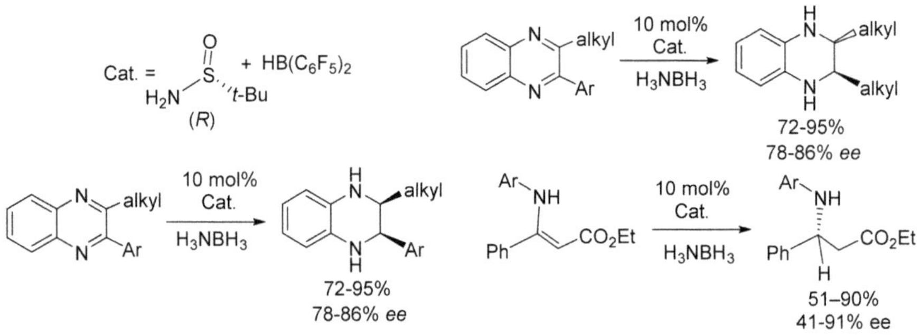

Scheme 6.13 Asymmetric transfer hydrogenations of 2-alkyl-3-aryl- and 2,3-dialkyl-quinoxaline substrates and *N*-substituted enamino esters.

(Scheme 6.12).[24] In contrast, 2,3-dialkylquinoxalines with enantiomeric excesses up to >99% were derived from the trans precursors. This same protocol was also applied to *N*-substituted enamino esters, affording β-amino acid derivatives in 51–90% yields with up to 91% enantiomeric excess (Scheme 6.13).[25]

6.5 Dehydrogenations

The transfer hydrogenations described in the previous two sections, clearly involve dehydrogenation of the source of dihydrogen. In this section, we focus on reactions in which FLPs mediated dehydrogenation where it is the oxidized products that are the center of interest.

In 2010, Miller and Bercaw[26] reported the use of the FLP, *t*-Bu₃P, and B(C₆F₅)₃ to effect the dehydrogenation of ammonia boranes, Me₂NHBH₃ and NH₃BH₃, affording the phosphonium-borate and the corresponding B/N dehydrocoupling products (Scheme 6.14). Also in 2010, the Manners group[27] demonstrated that silicon-based Lewis acids, such as Me₃SiOTf, in combination with sterically hindered amine or pyridine bases, behaved as FLPs. These systems also did not react directly with dihydrogen, although the combination of Me₃SiOTf and tetramethylpiperidine reacted with the amine–borane adduct Me₂NHBH₃ generating the salt [C₅H₆Me₄NH]OTf, Me₃SiH, and

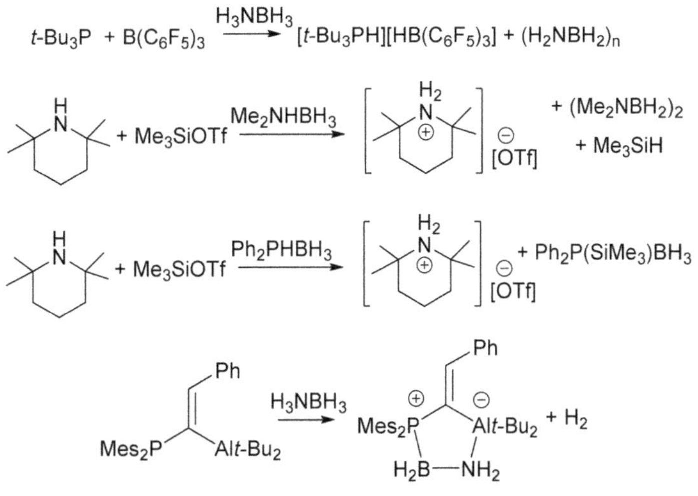

$$t\text{-Bu}_3\text{P} + \text{B(C}_6\text{F}_5)_3 \xrightarrow{\text{H}_3\text{NBH}_3} [t\text{-Bu}_3\text{PH}][\text{HB(C}_6\text{F}_5)_3] + (\text{H}_2\text{NBH}_2)_n$$

Scheme 6.14 Dehydrogenation of ammonia-boranes by FLPs.

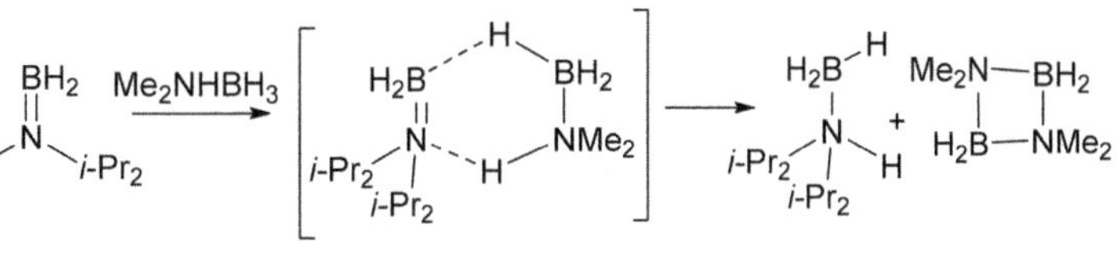

Scheme 6.15 Proposed reaction of i-Pr$_2$NBH$_2$ with Me$_2$HNBH$_3$.

the dehydrogenated product [Me$_2$NBH$_2$]$_2$ (Scheme 6.14). While such reactions can be viewed as dehydrogenation of amine-boranes, these reactions affect the net transfer of dihydrogen from amine-borane to the FLP. Similarly, the Sn(IV) Lewis acid Bu$_3$SnOTf in combination with 2,2,6,6-tetramethylpiperidine affected the analogous reaction although, again, this Sn/N FLP was unreactive with dihydrogen. Manners *et al.*[27] also showed that the corresponding reaction with the phosphine-borane Ph$_2$PHBH$_3$ with Me$_3$SiOTf and C$_5$H$_6$Me$_4$NH did not affect dehydrogenation but rather gave Ph$_2$P(SiMe$_3$)BH$_3$ and [C$_5$H$_6$Me$_4$NH$_2$][OTf] (Scheme 6.14). A subsequent paper by Appelt *et al.*[28] showed that the geminal FLP Mes$_2$PC=CHPh(Alt-Bu$_2$) affected the dehydrogenative capture of (H$_2$NBH$_2$) from ammonia-borane. In the case of Me$_2$NHBH$_3$, this FLP mediated dehydrogenation regenerated the FLP and generated the dimeric product (Me$_2$NBH$_2$)$_2$ (Scheme 6.14). The next year, Wass *et al.* extended the dehydrogenation of Me$_2$HNBH$_3$ to the FLP derived from the zirconocene species [Cp*$_2$Zr(ClC$_6$H$_5$)OC$_6$H$_4$Pt-Bu$_2$][B(C$_6$F$_5$)$_4$].[29]

A subsequent study by the Manners group[30] probed the mechanism of Me$_2$HNBH$_3$ dehydrogenation using the species i-Pr$_2$NBH$_2$. While not described as an FLP, this species is reminiscent of compounds of the form R$_2$PB(C$_6$F$_5$)$_2$. Moreover, kinetic study and computations supported a concerted mechanism for the dehydrogenation of ammonia borane, akin to the concerted activation of H$_2$ by FLPs (Scheme 6.15).

In 2016, Rivard and co-workers[31] showed that 2 mol% of the iminoborane (C_3H_2 $(N(i\text{-}Pr_2C_6H_3N)_2))NBR_2$ (R = Cl, Ph) affected the dehydrogenation of Me_2NHBH_3 at 70 °C. In the same year, 1 mol% of the xanthene-based intramolecular FLP (see Section 2.10) was used by Aldridge *et al.*[32] to dehydrogenate H_3NBH_3, $MeNH_2BH_3$ and Me_2NHBH_3. Subsequent reports showed the related reactivity of the geminal P/Ga species $Mes_2PC=CHPh(Gat\text{-}Bu_2)$,[33] the P/B FLP $C_6H_4(Pi\text{-}Pr_2)(B(C_6H_3(CF_3)_2)_2)$[34] and $t\text{-}Bu_2$ PCH_2BPh_2.[35] Recent computational papers have probed the potential of intramolecular B/N,[36] B/P,[37] and Pt/B[38] FLPs in ammonia-borane dehydrogenation. Most recently, the all metal-based FLP derived from the combination of the basic $[C_6H_4(2,6\text{-}(CH_2Pt\text{-}Bu_2)_2)$ PdH] and the acidic $CpWH(CO)_3$ was reported to catalyze the dehydrogenation of the Me_2NHBH_3 and $t\text{-}BuNH_2BH_3$, liberating dihydrogen.[39]

Manners *et al.*[40] have also exploited the Lewis acid $B(C_6F_5)_3$ to mediate the homocoupling of the P–H bonds of secondary or primary phosphines to generate P–P bonds and dihydrogen. Alternatively, the combination of tertiary silanes and phosphines can be dehydrocoupled to afford species of the form R_2PSiR_3 or $RP(SiR_3)_2$ (Scheme 6.16). The use of primary or secondary silanes or phosphines provides interesting cyclic Si–P dimers or trimers.

FLP dehydrogenation reactions have also provided access to unusual main group species. For example, the related reaction of the bulky stannane $(2,4,6\text{-}i\text{-}Pr_3C_6H_2)SnH_3$ with the FLP, $Pt\text{-}Bu_3/B(C_6F_5)_3$, in varying stoichiometries led to hydride abstraction and formation of $[(2,4,6\text{-}i\text{-}Pr_3C_6H_2)SnH_2(Pt\text{-}Bu_3)]^+$.[41] This species is unreactive toward dehydrogenation. However, excess FLP proved capable of dehydrogenating this species, affording the species $[(2,4,6\text{-}i\text{-}Pr_3C_6H_2)Sn(Pt\text{-}Bu_3)]^+$ containing the two-coordinate tin cation (Scheme 6.17).

$$PhPH_2 + H_3SiPh \xrightarrow{B(C_6F_5)_3} [PhPSiHPh]_3 + [PhPSiHPh]_2 + H_2$$

$$Ph_2PH + HSiEt_3 \xrightarrow{B(C_6F_5)_3} Ph_2PSiEt_3 + H_2$$

Scheme 6.16   Hetero- and homocoupling of PH bonds mediated by $B(C_6F_5)_3$.

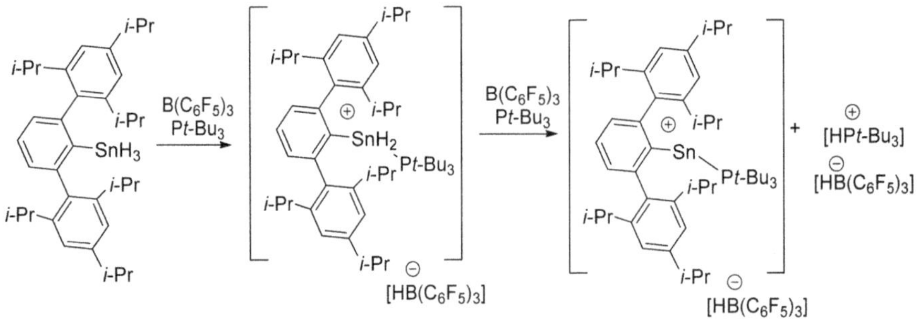

Scheme 6.17   Dehydrogenation of a tin hydride by an FLP.

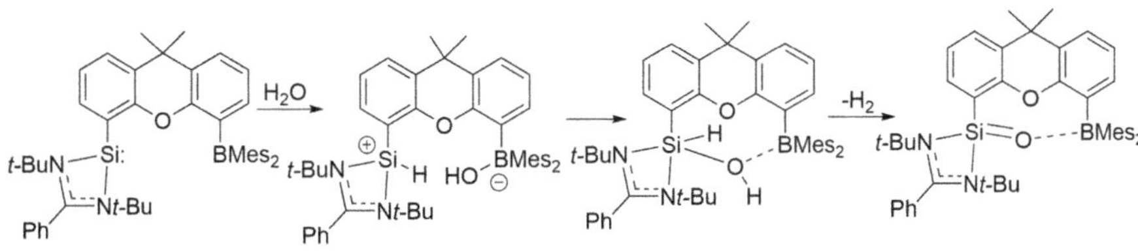

Scheme 6.18 Dehydrogenation of water by a Si/B FLP.

Another example of dehydrogenation followed the construction of the intramolecular silylene-borane FLP with a xanthene backbone, $(C_{13}H_8Me_2O)(B(Me_3C_6H_2)_2)(Si(t\text{-}BuN)_2CPh)$ by the Driess group.[42] This species exhibited unprecedented reactivity with H_2O affording the dehydrogenation of water, yielding the reactive silanone–borane product (Scheme 6.18). Computations revealed FLP activation of water was followed by hydroxide migration from boron to silicon and subsequent loss of dihydrogen. While the product silanone proved unstable with excess water, it was prepared independently from the initial FLP with N_2O.

6.6 CO Reduction

It became evident that FLPs could be used in a variety of organic reductions. At the same time, Fischer–Tropsch catalysis remained an important chemical problem typically limited to mediation by heterogeneous transition metal catalysts under generally forcing conditions.[43,44] The first report describing efforts to exploit FLP chemistry in this vein came from Bercaw and coworkers.[45] These authors described the reaction of the Re-complex with pendant boranes $[(BBNCH_2CH_2PPh_2)Re(CO)_4][BF_4]$ with dihydrogen in the presence of the base $((CH_2)_4N)_3PNt\text{-}Bu$. This combination resulted in the activation of dihydrogen promoting the reduction of a coordinated CO fragment affording $[(BBNCH_2CH_2PPh_2)Re(CO)_3CH_2O][((CH_2)_4N)_3PNHt\text{-}Bu]$ and $[((CH_2)_4N)_3PNHt\text{-}Bu][BF_4]$ (Scheme 6.19).

To probe stoichiometric metal-free CO reduction,[46] we employed the classic FLP, $B(C_6F_5)_3/t\text{-}Bu_3P$ in reaction with carbon monoxide and dihydrogen. At room temperature, the initial product was the formylborate salt $[t\text{-}Bu_3PH][(C_6F_5)_3BCHOB(C_6F_5)_3]$. Heating to 90 °C, this species underwent a rearrangement affording a migration of a C_6F_5 ring to the acyl carbon, providing $[t\text{-}Bu_3PH][(C_6F_5)_2BCH_2(C_6F_5)OB(C_6F_5)_3]$. Subsequent addition of carbon monoxide and dihydrogen gave the products $[t\text{-}Bu_3PH][(C_6F_5)BCH_2(C_6F_5)OB(C_6F_5)_3]$ and $[t\text{-}Bu_3PH][(C_6F_5)_2BCH(C_6F_5)O_2CB(C_6F_5)_3]$ in $2:3$ ratio respectively (Scheme 6.20).

In a related effort, the Erker group[47] has stabilized an unprecedented formylborane using an FLP. This $(\eta^2\text{–formylborane})$FLP complex, $C_7H_{10}(B(C_6F_5)_2)(PMes_2)(HC(O)B(C_6F_5)_2)$, reacted with dihydrogen, leading to the reduction of the formyl group, with cleavage of the C–O bond affording the methylene and hydroxy groups in $C_7H_{10}(B(C_6F_5)_2(\mu\text{-}OH)B(C_6F_5)_2)(CH_2PMes_2)$ (Scheme 6.21). This reaction is thought to proceed *via* dissociation of the B–O bond in the $(\eta^2\text{–formylborane})$FLP complex

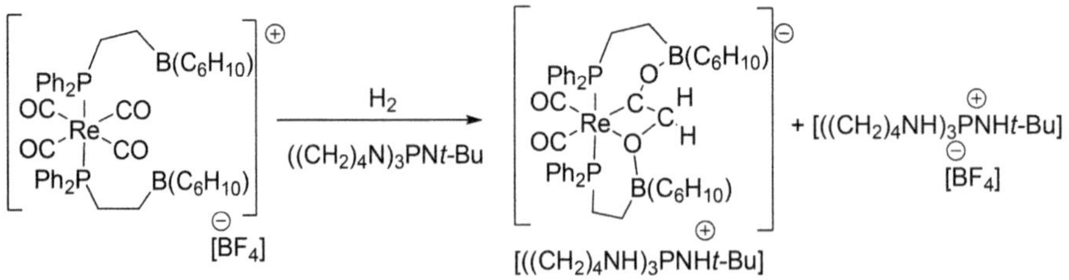

Scheme 6.19 Re-complex with pendant boranes with base and dihydrogen.

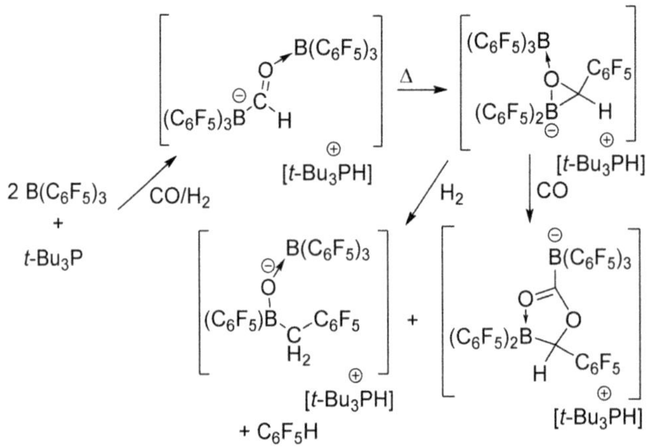

Scheme 6.20 Reactions of the classic FLP with CO and dihydrogen.

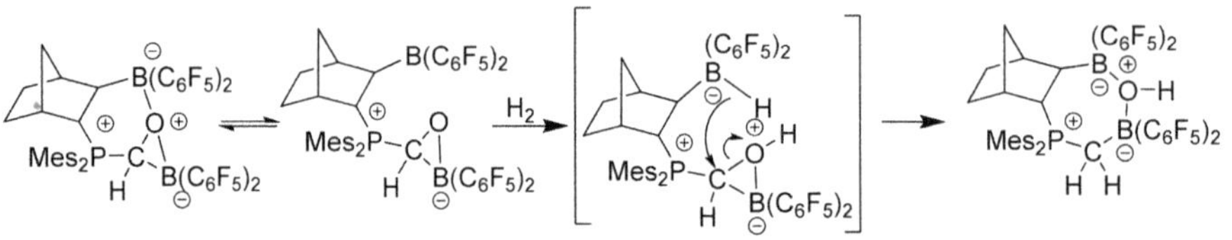

Scheme 6.21 Reduction of FLP-formyl borane complex.

allowing the dihydrogen activation between the boron and oxygen atom. Subsequent opening of the transient oxonium and proton migration afforded the observed product.

Expanding CO reduction chemistry to transition metal systems, the Erker group[48] also reported the formylhydridoborate complex hydride Cp*$_2$Zr(OMes)O=CHBH (C$_6$F$_5$)$_2$ derived from the reaction of the bulky metallocene hydride Cp*$_2$Zr(H)OMes with HB(C$_6$F$_5$)$_2$ and CO. This species was dynamic in solution *via* an endergonic dissociation of the boron from the formyl fragment, affording access to B/O FLP. Reaction with dihydrogen proceeded with the generation of Cp*$_2$Zr(OH)OMes and MeB(C$_6$F$_5$)$_2$, which slowly reacted to further yield Cp*$_2$Zr(OMes)O(H)B(H)(C$_6$F$_5$)$_2$ and methane (Scheme 6.22).

In a very recent effort,[49] we have employed alkali-metal amides as FLPs (see Section 5.16) in reactions with CO and dihydrogen. Cy$_2$NLi under CO alone was shown to affect CO homologation. This, in addition to the previously described ability to activate dihydrogen (see Section 5.16) prompted an examination of reactions of Cy$_2$NLi with syn-gas. This afforded the product [Cy$_2$NCH$_2$OLi]$_6$ albeit in low yield, in addition to the species Cy$_2$NC(O)CH$_2$OH after 48 hours (Scheme 6.23). The formation of the latter species was

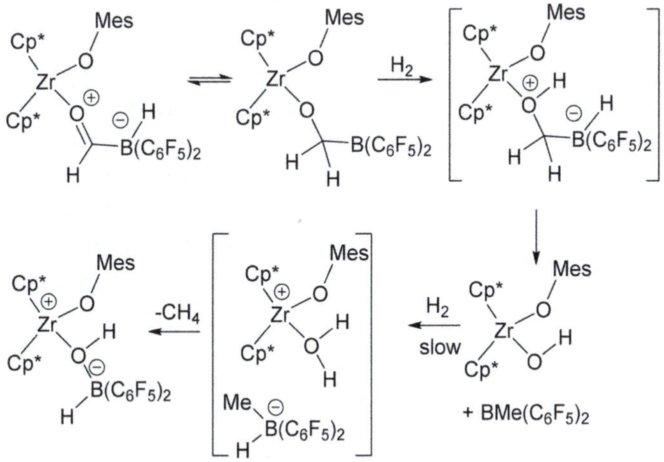

Scheme 6.22 Reaction of Zr-formylhydridoborate complex with dihydrogen *via* a B/O FLP.

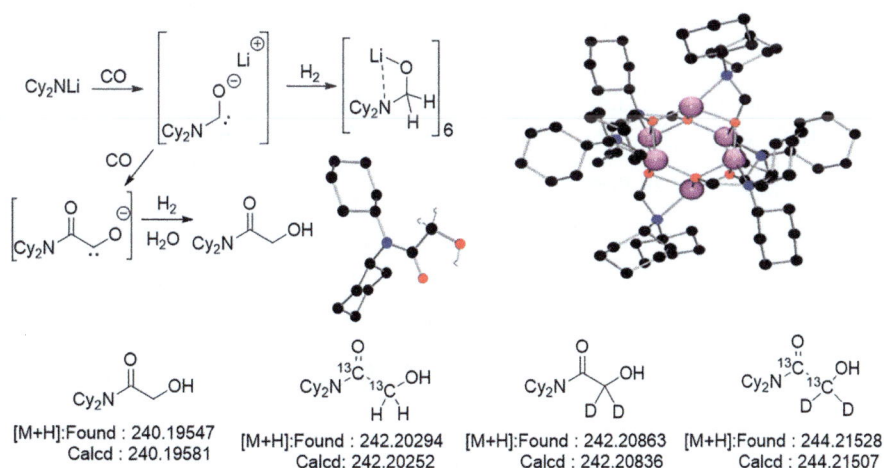

Cy$_2$N \ OH	Cy$_2$N ^{13}C ^{13}C OH H H	Cy$_2$N OH D D	Cy$_2$N ^{13}C ^{13}C OH D D
[M+H]:Found : 240.19547 Calcd : 240.19581	[M+H]:Found : 242.20294 Calcd : 242.20252	[M+H]:Found : 242.20863 Calcd : 242.20836	[M+H]:Found : 244.21528 Calcd : 244.21507

Scheme 6.23 Reaction of Cy$_2$NLi with CO/dihydrogen; POV-ray depictions of [Cy$_2$NCH$_2$OLi]$_6$ and Cy$_2$NC(O)CH$_2$OH; HRMS data for isotopomers of Cy$_2$NC(O)CH$_2$OH.

affirmed by crystallography, as well as the preparation of the isotopomers derived from ^{13}CO and/or D_2 incorporation (Scheme 6.23). These products are consistent with the initial generation of a carbene-type intermediate which reacts with either CO or dihydrogen, affecting C–C and C–H bond formation, which are reactions fundamental to Fischer–Tropsch chemistry (Scheme 6.23).

Computations showed that the reactions of $[(THF)Li(\mu\text{-}NCy_2)]_2$ with CO proceed *via* a five-membered ring transition state, involving nucleophilic attack at the carbonyl-carbon affording a 'bent anionic carbene-like' fragment. Subsequent and sequential reactions with CO afford an avenue to CO homologation with a relatively low energy barrier for each step. The reaction of each of these intermediates with dihydrogen was also computed, displaying activation barriers that are slightly higher than those with CO, but still accessible at room temperature (Scheme 6.24). These findings infer that the product mixture will be dependent on the availability and solubility of CO and dihydrogen in the reaction solvent, thus accounting for the complex mixture of products derived from the fundamental reactions of the Fischer–Tropsch process.

The above reactions demonstrate the viability of FLPs in the stoichiometric reduction reactions of CO as well as in basic reactions of Fischer–Tropsch chemistry. While this is interesting from a fundamental perspective, efforts to exploit these findings for catalysis will have their challenges. In the case of Fischer–Tropsch reactivity, it will be

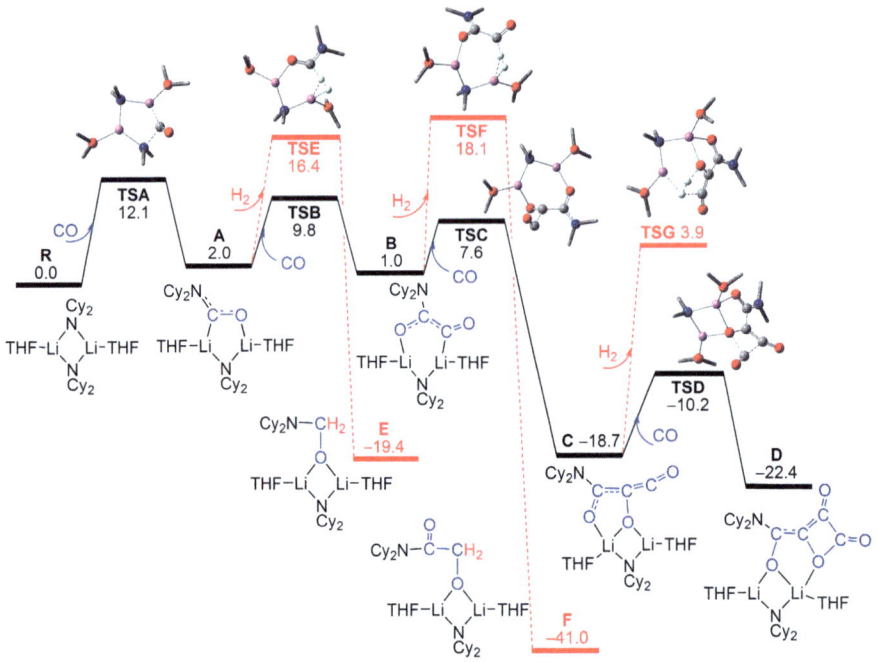

Scheme 6.24 Computed reaction Gibbs free energy profile for the reaction of $[(THF)Li(\mu\text{-}NCy_2)]_2$ with CO and dihydrogen (energies in kcal mol^{-1}, at 298 K and 1 mol L^{-1}).[49] Reproduced from ref. 49 with permission from American Chemical Society, Copyright 2021.

necessary to find an FLP that balances the energetics of hydrogen activation and CO homologation to permit these two reactions to occur in a sequential fashion, affording access to higher molecular weight hydrocarbons. Nonetheless, these initial findings demonstrate that the notion of a transition metal-free Fischer–Tropsch process is worthy of further exploration.

6.7 CO_2 Reduction

The reduction of CO_2 has been recognized as a chemical transformation worthy of much attention in recent years. While this offers a unique approach to the valorization of a 'waste' material, the reduction of CO_2 on a large scale would also represent an alternative energy source to fossil fuels thereby providing strategies to address global warming concerns. As early collaborative studies with the Erker group demonstrated, the ability of FLPs to capture CO_2[50] and the use of FLPs in the potential reduction of CO_2 has garnered some attention.

It was the work of Ashley and O'Hare[51] in 2009, that first demonstrated the potential of such systems to affect the reduction of CO_2. In this seminal work, a 1:1 mixture of tetramethylpiperidine and $B(C_6F_5)_3$ in toluene under a CO_2/dihydrogen atmosphere was heated to 160 °C for 6 days, resulting in the quantitative production of $MeOB(C_6F_5)_2$ (Scheme 6.25). A subsequent distillation provided MeOH in 17–25% yield.

In subsequent work,[52] O'Hare *et al.* showed that the FLP derived from $(C_6F_4(C_6F_5))_3B$ and Pt-Bu_3 activated dihydrogen and that the resulting salt reacted with CO_2 to give the formate salt $[HPt$-$Bu_3][(C_6F_4(C_6F_5))_3BO_2CH]$, which is thought to be the first step in the reduction of CO_2 by an FLP. A related study showed analogous reactivity of lutidine and $B(C_6F_5)_3$[53] affording $[C_5H_3Me_2NH][HCO_2B(C_6F_5)_3]$, while computational studies have addressed the forcing conditions required for CO_2 reduction,[54] as well as CO_2 reduction by intramolecular FLPs.[55,56] Exploiting amine/borane FLPs, Wang *et al.*[57] showed that the salt $[H_2NR_2][R_2NCO_2BH(C_6H_2(CF_3)_3)_2]$, derived from the reaction of HNR_2 (R = Et, i-Pr), $HB(C_6H_2(CF_3)_3)_2$ and CO_2, lost dihydrogen to generate the boronic ester R_2NCO_2 $B(C_6H_2(CF_3)_3)_2$. Upon exposure to 50 bar of dihydrogen at 70 °C, this latter product (R = i-Pr) was converted to a mixture of products including $R_2NC(H)O(HB(C_6H_2(CF_3)_3)_2)$, $HOB(C_6H_2(CF_3)_3)_2$, and $[H_2NR_2][HCO_2BH(C_6H_2(CF_3)_3)_2]$ (Scheme 6.26). This reaction was proposed to proceed *via* the B/O FLP activation of dihydrogen with subsequent proton migration to nitrogen.

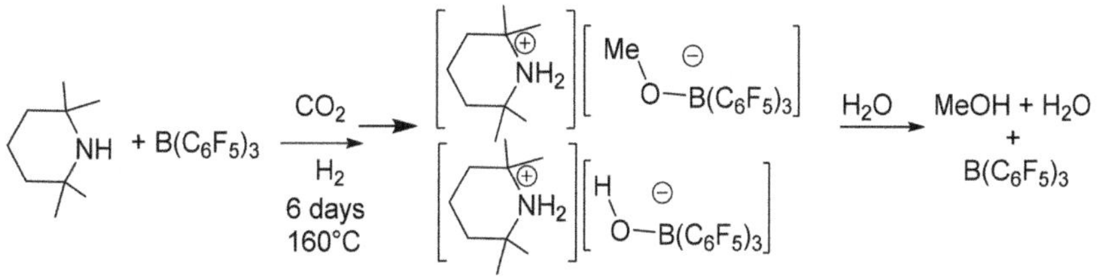

Scheme 6.25 Stoichiometric FLP reduction of CO_2.

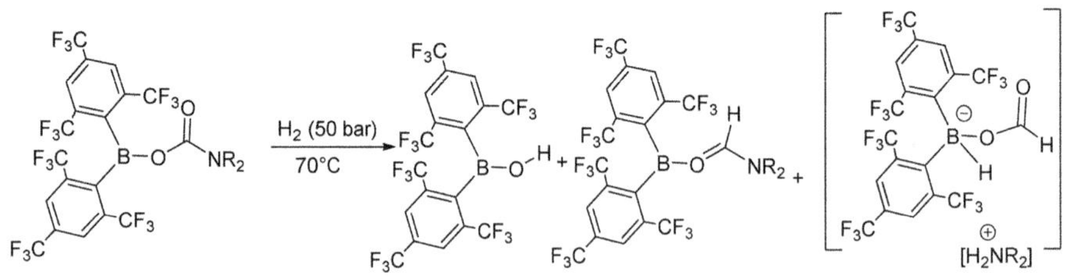

Scheme 6.26   Reaction of the boron–carbamate with dihydrogen.

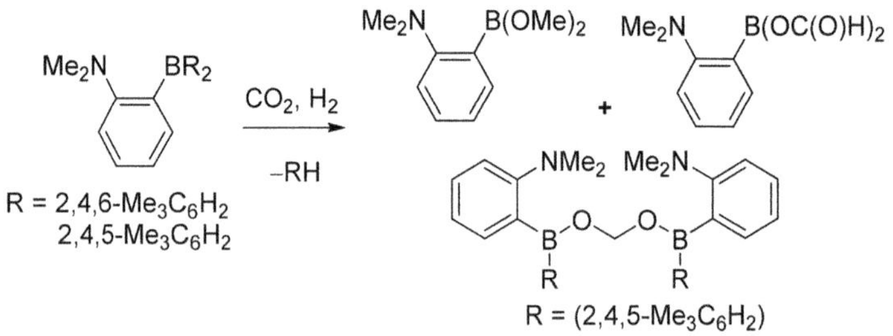

Scheme 6.27   Hydrogenation of CO_2 with $[(R_2B)(Me_2N)C_6H_4]_2$.

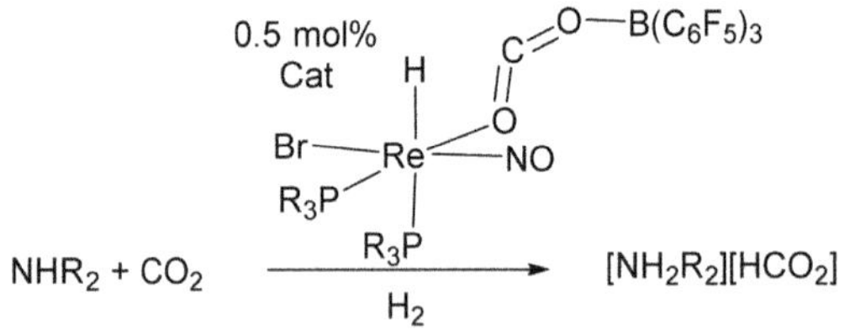

Scheme 6.28   ReH/borane FLP reduction of CO_2.

   While several studies of CO_2 capture and reduction using silane or boranes have been reviewed,[58] the reduction of CO_2 with dihydrogen has received lesser attention. In 2015, in a collaborative effort with the Fontaine group,[59] we reported that the intramolecular FLPs $(R_2B)(Me_2N)C_6H_4$ (R = Mes, $2,4,5\text{-}Me_3C_6H_2$) reacted with dihydrogen followed by protodeborylation reactions to give the dimeric species $[(H_2B)(Me_2N)C_6H_4]_2$. Nonetheless, these amino-borane FLPs reacted with CO_2 and dihydrogen, affording the formyl and methoxy derivatives $(R_2B)(Me_2N)C_6H_4$ (R = C(O)H, MeO) as well as the acetal $CH_2((O(R)B)(Me_2N)C_6H_4)_2$ (Scheme 6.27).
   Transition metal-based Lewis acids have also been exploited for the hydrogenation of CO_2. For example, in 2013, Berke and coworkers[60] studied the reaction of the Re complex $ReHBr(NO)(PR_3)_2$ with $B(C_6F_5)_3$, demonstrating the FLP capture of CO_2 (Scheme 6.28). Hydrogenations of CO_2 using 0.5 mol% of the rhenium complex and $B(C_6F_5)_3$ as

the catalyst in the presence of $C_5H_6Me_4NH$ yielded $[C_5H_6Me_4NH_2][HCO_2]$ with a turnover frequency of up to 7.5 h^{-1}. Using various bases these authors demonstrated that the steric bulk and the basicity of the base were critical in affecting CO_2 reduction.

For example, the complex $[MeC(CH_2PPh_2)_3Cu(NCMe)][PF_6]$[61,62] was used in the presence of $C_{10}H_{16}N_2$ (DBU) to catalytically convert dihydrogen and CO_2 to the formate salt, $[C_{10}H_{17}N_2][HCO_2]$ (Scheme 6.29). The mechanism was thought to involve initial heterolytic cleavage of dihydrogen affording a transient Cu-hydride and protonated base. Subsequent reaction of the Cu–H fragment with CO_2 afforded the formate anion. The efficiency of such transformations was recently improved[63] using the bis-amidine complex $[((C_3H_4MeN_2)_2CMe_2)Cu(O_2CMe)_2]$ and 20 bar CO_2, 40 bar dihydrogen at 100 °C for 21 h. This latter system led to a turnover number of 1600. Very recently, 1 mol% of the phosphaalkene pincer Cu(I) cations $[C_5H_3N(CHPR)_2Cu]^+$ (R = 2,4,6-t-Bu$_3$C$_6$H$_2$, $C_6H(C_3H_2Et_4)_2$) were used a Lewis acid catalyst in the conversion of the base $C_{10}H_{16}N_2$ (DBU) under dihydrogen and CO_2 at 90 °C to the salt $[C_{10}H_{17}N_2][HCO_2]$ after 24 h (Scheme 6.29).[64] A computational study also explored the potential of the Co and Fe complexes $[N(CH_2CH_2Pi\text{-}Pr_2)_2Co(C(O)CH_2C_5H_3NO)]^{+}$[65] and $[(t\text{-}BuNCH_2CH_2Pt\text{-}Bu)_2Fe(CO)(CN)_2]$[66] to mediate the reaction of CO_2 and dihydrogen to formic acid.

In an interesting hybrid approach, Bertrand and coworkers[67] reported the use of the Cu(I) species $((2,6,\text{-}i\text{-}Pr_2C_6H_3)NC_4H_2Me_2Et_2)Cu(H_2BH_2))$ with the FLP comprised of an N-base/$B(C_6F_5)_3$ (Scheme 6.30). This system acted as a tandem catalyst, as it mediated the reduction of CO_2 to the corresponding formate salt. The use of the FLP served to accelerate the activation of dihydrogen. The optimal reaction conditions employed DBU as the base, 15 bar CO_2, 45 bar dihydrogen, at 100 °C and a catalyst loading of 0.025%, affording a turnover number of 1881 after 24 h. This combination tandem catalyst system outperformed both the Cu(I) systems described above as well as the FLP system alone.

Targeting applications in organic synthesis and exploiting the potential of transfer hydrogenation for the reduction of CO_2, Bhanage and coworkers[68] used $B(C_6F_5)_3$ and Me_2NHBH_3 to mediate the N-formylation of amines using CO_2 at 80 °C. This was

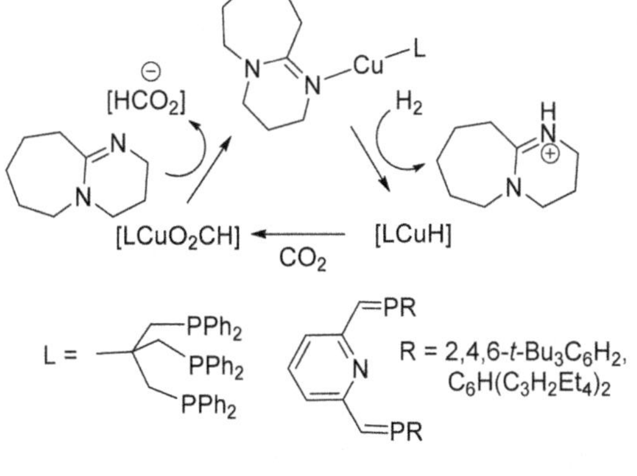

Scheme 6.29 Proposed mechanism of CO_2 reduction by Cu(I) species and base.

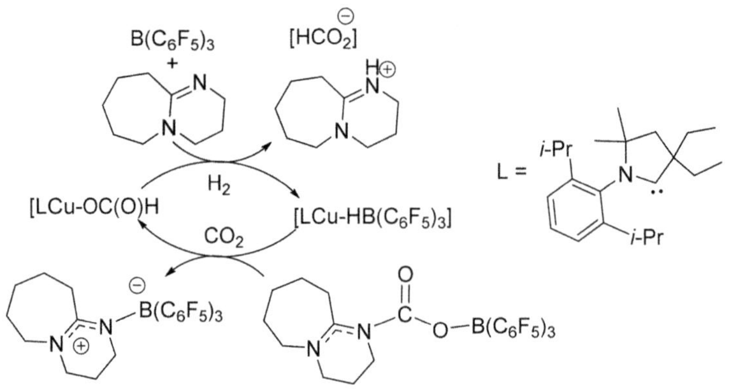

Scheme 6.30 Cu(I)/FLP tandem catalyst for the reduction of CO_2.

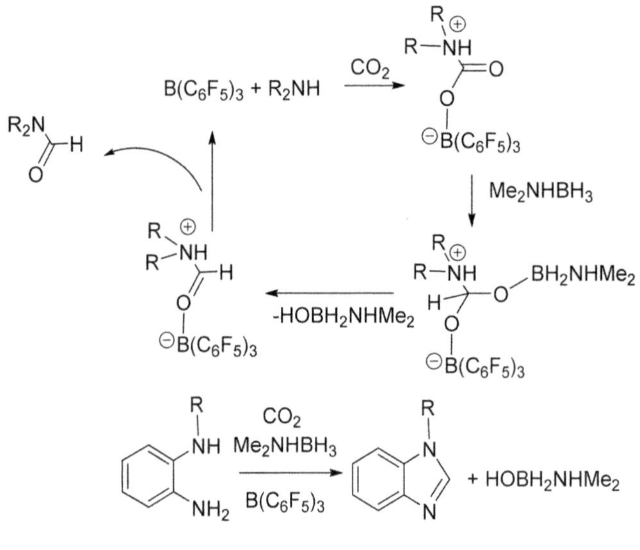

Scheme 6.31 FLP mediated transfer hydrogenation/formylation of amines.

applied to efficiently prepare a range of 18 formylated products, generally in good yields under relatively mild conditions. In addition, the corresponding reactions of o-phenylenediamines provided facile access to five substituted benzimidazoles. The proposed mechanism involved the amine and the borane acting as an FLP to capture CO_2, followed by reaction with Me_2NHBH_3 and loss of Me_2NHBH_2OH to give the N-formylated amine and liberating $B(C_6F_5)_3$ for further reaction (Scheme 6.31).

The above systems demonstrate a series of FLP approaches to the reduction of CO_2. While the direct reduction to methanol has proven feasible, the exploitation of FLPs in tandem catalysis or applications targeting organic syntheses seems to offer the greatest potential for future development.

6.8 Oxidation of Dihydrogen

In an innovative application of FLP chemistry, the Ashley and Wildgoose groups[69] collaborated in the exploration of the utility of the classic FLP in dihydrogen oxidation. To this end, the combination of t-Bu$_3$P and B(C$_6$F$_5$)$_3$ in the activation of dihydrogen was studied together with nonaqueous electrochemical oxidation of the resulting borohydride (Scheme 6.32). This approach resulted in the net oxidation of dihydrogen to two protons and two electrons. The use of FLPs in this fashion reduced the potential required for this oxidation by 610 mV at a glassy carbon electrode (GCE). A subsequent study[70] explored the use of a Pt electrode and found a 390 mV reduction in the oxidation potential compared to the GCE. A further extension of this work,[71] used the FLP derived from t-Bu$_3$P and the NHC-stabilized borenium cation [(C$_3$H$_2$(N(i-Pr$_2$C$_6$H$_3$))$_2$)(BC$_8$H$_{14}$)]$^+$ to effect dihydrogen oxidation. This was shown to further decrease the required voltage by 910 mV. The latter FLP system provided improved catalyst stability and recyclability.

Wildgoose and Ingleson[72] collaborated to further expand on the potential of FLP systems for dihydrogen oxidation. Exploiting a combination of the carbon-based N-methylacridinium cation and a borane, these authors demonstrated the synergy of fast hydrogen cleavage kinetics and excellent electrochemical behavior attributed to a 'hydride shuttle'. This novel combined system decreased the required potential for hydrogen oxidation by 1 V and was recycled multiple times.

A more recent collaborative[73] effort among the Wildgoose, Ashley and Slootweg groups developed an alternative approach to dihydrogen oxidation. These teams employed a borane in combination with a chemical reductant to generate a radical anion that reacted with dihydrogen. Thus, using B(C$_6$(NO$_2$)$_2$Me$_3$)$_3$ or BMes$_3$ in combination with Cp*$_2$Co or Na, prompted the homolytic cleavage of dihydrogen. While this reaction was slow, allowing the characterization of intermediates, this work provided a new approach exploiting radicals in dihydrogen oxidation.

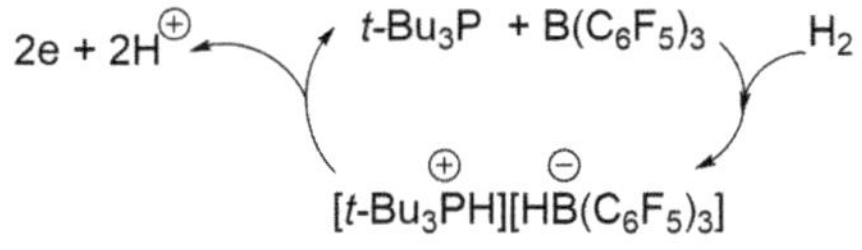

Scheme 6.32 FLP/electrochemical oxidation of dihydrogen.

6.9 Main Group FLP Hydrogenation

While FLP hydrogenations of organic substrates have been the focus of major research efforts, the utility of FLPs for main group reductions offers an avenue to broaden the applications of FLP chemistry. One possibility that we targeted early on was the stoichiometric reduction of P–P sigma bonds. To this end, P_5Ph_5 was reacted with $B(C_6F_5)_3$ and dihydrogen, which led to the generation of the phosphine–borane adduct $(PhPH_2)B(C_6F_5)_3$ in nearly quantitative yield.[74] In contrast, $(t-Bu_2P)_2$ in the presence of $B(C_6F_5)_3$ and dihydrogen afforded the salt $[t-Bu_2PPHt-Bu_2][HB(C_6F_5)_3]$ (Scheme 6.33).[75]

In a subsequent study,[76] as the borane $B(C_6F_4H)_3$ is not susceptible to *para*-attack, it was used to catalyze the dehydrocoupling of R_2PH (R = Ph, *p*-tol) to P_2R_4 and $PhPH_2$ to $(PhP)_5$ (Scheme 6.34). In addition, the secondary phosphines were dehydrocoupled with concurrent hydrogenation of either the silyl enol ether $MeC(OSiMe_3)=CH_2$ or the imine $PhCH=Nt-Bu$. In the case of Ph_2PH, the proposed mechanism for dehydrocoupling involved the formation of a transient salt of the form $[Ph_2(H)PPPh_2][HB(C_6F_4H)_3]$ (Scheme 6.34).

In more recent work, a collaboration with the Paradies group[77] exploited the Lewis acidity of halophosphonium cations to mediate the reduction of phosphine oxides. To this end, phosphine oxides were treated with oxalyl chloride under 80 bar of dihydrogen at 80 °C. This prompted the generation of the corresponding dichlorophosphorane. Dissociation of chloride generated the Lewis acidic chlorophosphonium cation. This species, in combination with phosphine oxide acts, as an FLP to activate dihydrogen (Scheme 6.35), ultimately leading to the reduction of the phosphine

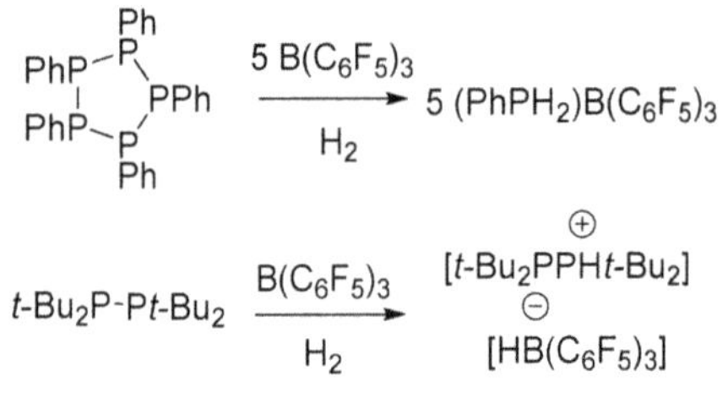

Scheme 6.33 Borane/polyphosphine reactions with dihydrogen.

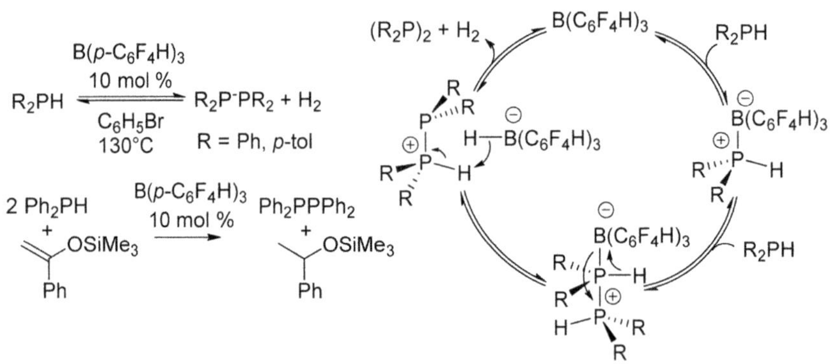

Scheme 6.34 Catalytic dehydrocoupling of phosphines and transfer hydrogenation.

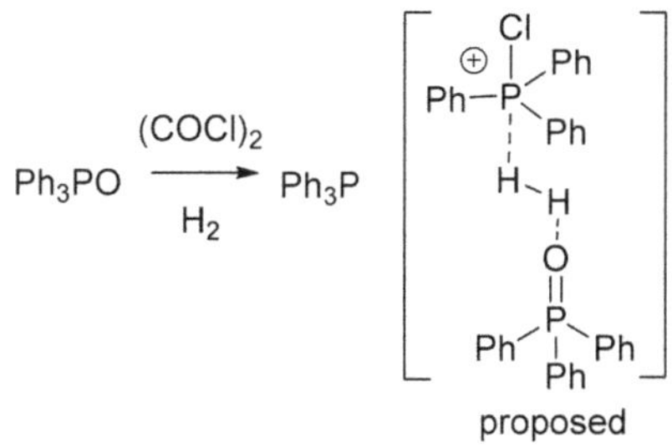

$$Ph_3PO \xrightarrow[H_2]{(COCl)_2} Ph_3P$$

$$\left[\begin{array}{c} \overset{Cl}{\underset{\oplus}{\underset{Ph-P}{\vert}}} \overset{Ph}{\underset{Ph}{\diagdown}} \\ \overset{H}{\underset{\diagdown}{\vert}} \underset{H}{} \\ \overset{\vert}{\underset{O}{}} \\ \overset{O}{\underset{\parallel}{}} \\ Ph \overset{P}{\underset{\vert}{\diagup}} \overset{\diagdown}{} Ph \\ Ph \end{array} \right]$$

proposed

Scheme 6.35 *In situ* generation of P$^+$/O FLP affording phosphine oxide reduction.

oxide to the corresponding phosphine. Thus, this protocol which generated an *in situ* FLP proved effective for a range of triarylphosphine oxides, providing the product phosphines in yields ranging from 48–94%. It was, however, ineffective for trialkyl phosphine oxides. A subsequent detailed computational study by Grimme *et al.*[78] suggested that the use of the borane $B(2,6-F_2C_6H_3)_3$ lowers the activation barrier, as the combination of the borane with chloride or Ph_3P provides an energetically favorable avenue to dihydrogen activation. This finding suggested that this reduction of phosphine oxides operates *via* an autocatalytic mechanism.

6.10 N$_2$ Reduction?

Ever since uncovering the ability of FLPs to activate dihydrogen, the potential for applications in the Haber–Bosch reduction of dinitrogen was viewed as an exciting 'holy grail'. However, the interactions of dinitrogen with non-metal elements were limited. The binding of dinitrogen to BF_3 had been studied computationally, and the species $(N_2)BF_3$ was observed spectroscopically *via* supersonic expansion at 1 bar and 170 K.[79] Interestingly, a computational study of the species $Ph_3PNNPPh_3$[80] controversially described this as dinitrogen stabilized by the donation from two phosphine donors,[81] although this species is not generated in this fashion.

In our efforts, we noted that diazomethanes lost dinitrogen to generate a transient carbene,[82] and thus we probed the synthesis of a diazomethane adduct of $B(C_6F_5)_3$. Initial efforts revealed the loss of dinitrogen and the insertion of the carbene fragment into B–C bonds (Scheme 6.36).[83] However, the use of bulkier diazomethanes slowed the loss of dinitrogen allowing the isolation of $Ph_2CN_2B(C_6F_5)_3$ (Scheme 6.36). This species showed the end-on binding to boron; however, it is noteworthy that this species was extremely sensitive, losing dinitrogen above −40 °C.[84] Nonetheless, this species was stabilized by subsequent single electron transfer reactions. This was achieved with the addition of a metallocene to the diazomethane adduct. The transient radical species underwent C–H bond activation and afforded stable, isolable products (Scheme 6.36).[85]

The above observations suggested the possibility of dinitrogen capture by the cooperative action of an electron donor and an acceptor. Efforts to employ carbenes and boranes to this end were not successful. Nonetheless, the notion was conceptually supported by the subsequent chemistry reported by Braunschweig *et al.*[86,87] In this landmark work, dinitrogen reacted with a boron(I) species, providing the first example of

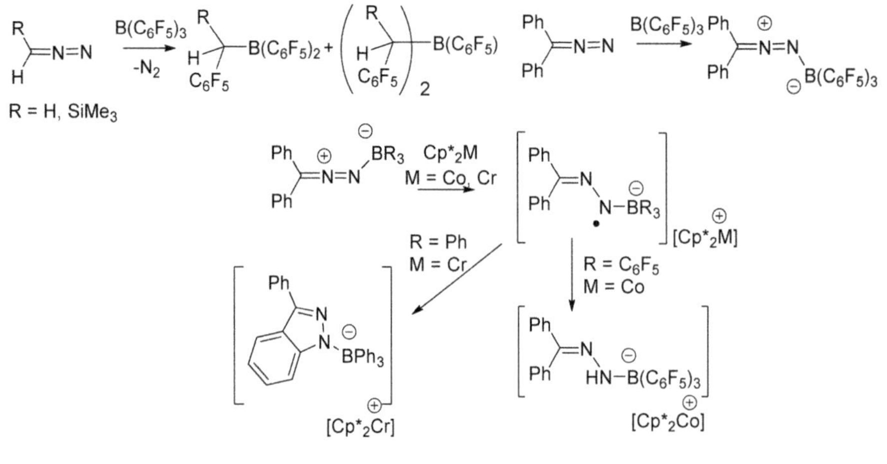

Scheme 6.36 Reactions of diazomethanes with boranes.

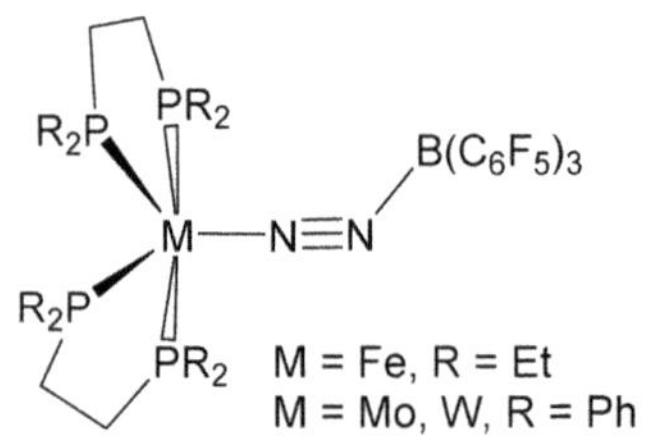

Figure 6.1 Interactions of metal-bound dinitrogen with borane.

direct dinitrogen activation by a non-metal species. Indeed, the boron(I) is isoelectronic to a carbene, with both electron acceptor and donor orbitals residing on the boron atom. Moreover, from the perspective of FLP chemistry, it was noted that B(I) species can be viewed as B(III) plus two electrons, thus furthering speculation that a suitably designed FLP could mimic this reactivity. This however has not been achieved yet.

Other efforts to exploit an FLP approach to the reduction of dinitrogen have been reported. The groups of Szymczak[88] and Simonneau[89] both used $B(C_6F_5)_3$ in combination with the metal–dinitrogen complexes $(Et_2PCH_2CH_2PEt_2)_2FeN_2$ and $(Ph_2PCH_2CH_2PPh_2)_2MN_2$ (M = Mo, Cr), respectively (Figure 6.1). Szymczak et al.[88] showed that interaction of the Lewis acid with the metal-bound nitrogen facilitated protonation, while Simonneau[89] demonstrated borylation and silylation of the dinitrogen fragment. A related report described the analogous tungsten chemistry[90] while the related Ru-based-dinitrogen chemistry was addressed in a computational study.[91]

While the above studies demonstrate the ability of FLP systems to activate nitrogen for further reaction, it was the Liddle group that took this a step further towards FLP hydrogenation of dinitrogen.[92] These authors generated a Ti_2Mg_2-nitride species from the reduction of a tris-amidoTiCl precursor. The nitride cluster formulated as $[N(CH_2CH_2N(SiMe_3))_3MgNTi]_2$ was shown to react with dihydrogen in the presence of t-Bu$_3$P and B(C$_6$F$_5$)$_3$, liberating NH$_3$ and generating a product formulated as [N(CH$_2$CH$_2$N(SiMe$_3$))$_3$MgTi]$_2$[HB(C$_6$F$_5$)$_3$]$_6$ (Scheme 6.37). This reaction was thought to involve FLP activation of dihydrogen and subsequent protonation of the nitride, providing an avenue to ammonia. This interesting finding illustrated the ability of FLP hydrogenations to participate in nitrogen reduction.

In a more recent development, the Liddle group[93] has also reported on the reaction of the uranyl-nitride with boranes. In the presence of BPh$_3$, the adduct K[N(CH$_2$CH$_2$N(Si i-Pr$_3$))$_3$UNBPh$_3$] was formed; however, the use of BMes$_3$ generated a B/N FLP. This latter combination reacted with dihydrogen, and led to the protonation of the nitride, ultimately affording N(CH$_2$CH$_2$N(Sii-Pr$_3$))$_3$UNH$_2$ and the radical [BMes$_3$]$^{\cdot-}$. The species N(CH$_2$CH$_2$N(Sii-Pr$_3$))$_3$UNH$_2$ formed a weak adduct with BPh$_3$ but reacted with dihydrogen, providing K[N(CH$_2$CH$_2$N(Sii-Pr$_3$))$_3$UH with the liberation of the ammonia adduct [H$_3$NBPh$_3$]. The species N(CH$_2$CH$_2$N(Sii-Pr$_3$))$_3$UNH$_2$ was also shown to react with Me$_3$SiCl to form Me$_3$SiNH$_2$ and N(CH$_2$CH$_2$N(Sii-Pr$_3$))$_3$UCl. Subsequent reaction with sodium azide, regenerated the initial nitride complex (Scheme 6.38). While this sequence of reactions in principle provided a cyclic process in which the nitride complex liberated ammonia with the ultimate regeneration of nitride species, a fully catalytic cycle remains a target for future development.

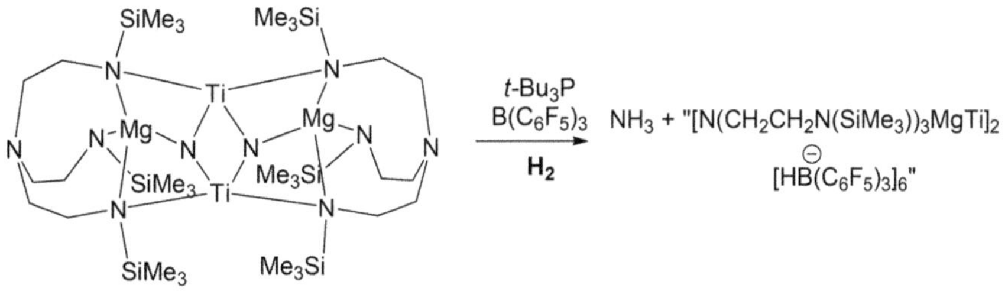

Scheme 6.37 FLP reduction of Ti$_2$Mg$_2$-nitride species.

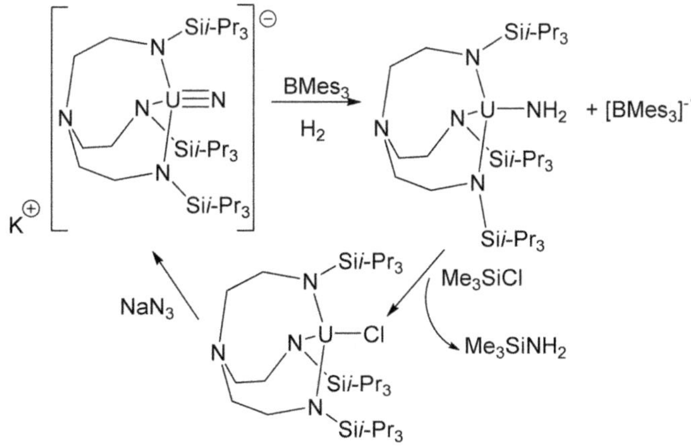

Scheme 6.38 Reactions of uranium nitride with borane and dihydrogen and regeneration of nitride species.

The chemistry described in this section provides support for the possibility of an FLP-mediated nitrogen reduction. While the presented data suggest the potential of judiciously selected FLPs to capture dinitrogen, this has not yet been achieved. Nonetheless, the known work does illustrate that FLPs can activate and reduce captured dinitrogen fragments. This leaves us tantalized by the prospects of future efforts.

6.11 Enzyme Models

The activation of dihydrogen by FLPs also has relevance to the activation of dihydrogen by enzymatic systems. In the case of [Fe]-hydrogenase, this enzyme was initially thought to be an organic hydrogenation catalyst but was subsequently shown to involve a Fe-guanylylpyridinol cofactor.[94] This enzyme mediates the reduction of methenyltetrahydromethanopterin to methylene tetrahydromethanopterin by delivering hydride to the imidazolium acceptor (Scheme 6.39).

In an ingenious approach, Meyer and coworkers[95] employed an FLP consisting of a Lewis basic Ru-metalate of the form $[CpRu(CO)_2]^-$ with the imidazolinium salt $[(CH_2NC_6F_2H_3)_2C(C_6H_4Me)]^+$. In this case, the Ru species acted as the basic site, while the imidazolium acted as the Lewis acid. This combination reacted with dihydrogen, prompting hydride delivery to the imidazolium and protonation of the Ru center (Scheme 6.40). In this fashion, this FLP mimics the enzymatic reactivity of hydrogenase.

It is also important to note that the active site of two other related hydrogenase enzymes have also been found to contain bimetallic active sites and thus are identified as [NiFe]- and [FeFe]-hydrogenase.[96] In the latter case, proximal to the bimetallic site is

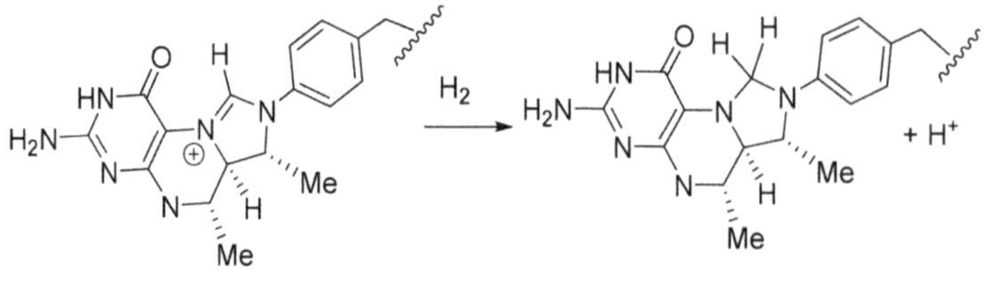

Scheme 6.39 Reduction of methenyltetrahydromethanopterin to methylene tetrahydromethanopterin.

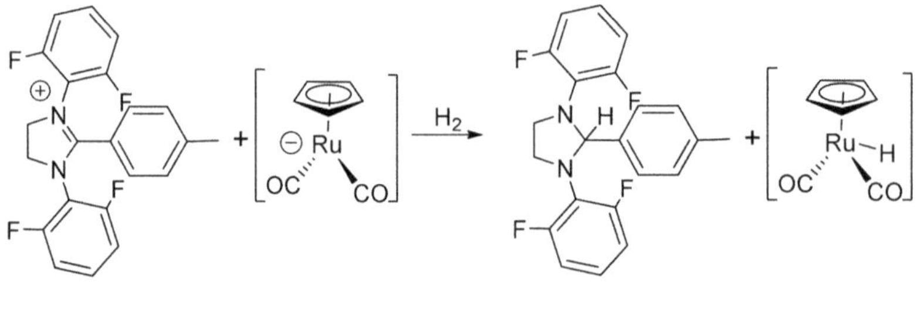

Scheme 6.40 Activation of dihydrogen by a Ru(0)/imidazolium FLP.

an amine base. Models developed by Bullock and coworkers have been described above (see Section 5.13). These illustrate an FLP-type activation of dihydrogen *via* protonation of the base and hydride delivery to the metal center. These data further demonstrate that the implications of the concept of FLP chemistry extend well beyond main group chemistry.

6.12 Solid-state FLPs

In 2018, Erker and collaborators[97] demonstrated an interesting contrast between solution and solid-state reactions. This team showed that mixtures of the Lewis acid and base $PCy_3/B(C_6F_5)_3$ in solution rapidly reacted by nucleophilic aromatic substitution to give the zwitterionic species, $[Cy_3PC_6F_4BF(C_6F_5)_2]$. In contrast, in the solid state, such deactivation is suppressed, allowing the solid FLP to react with 50 bar dihydrogen as revealed by solid-state NMR spectroscopy after 10 days. Analogous behavior was also demonstrated by the phosphines PCy_2Ph and PPh_2t-Bu. This solid-state reactivity was also observed after 10 h of reaction of a suspension of the FLP in perfluoromethylcyclohexane at ambient temperatures using 1.5 bar of dihydrogen. While the precise details of the mechanism of reaction in the solid-state remain the subject of speculation, these findings revealed that the cleavage of dihydrogen can be achieved even with combinations of Lewis acids and bases that do not behave as FLPs in solution. Similarly, the reactivity of an intramolecular B/N FLP with dihydrogen was also documented in the solid state.[98]

6.13 Supported FLPs

In 2016, O'Hare and coworkers[99] reacted solid-phase, silica-supported Lewis acid of the form $R^1R^2R^3SiOB(C_6F_5)_2$ with the Lewis base t-Bu$_3$P. A suspension of the resulting adduct under 2 bar of dihydrogen at 65 °C resulted in the cleavage of dihydrogen (Scheme 6.41). This same reactivity was mimicked using the soluble silica analog hepta-siloxane species $(RSi)_8O_{10}(O_2B(C_6F_5)_2)$ and t-Bu$_3$P.

Exploiting another strategy to a supported FLP catalyst, Taoufik *et al.*[100] used $HOC_6H_4PPh_2$ in reaction with silica treated with Ali-Bu$_3$ (Scheme 6.42). This provided an avenue to the grafting of the phosphine onto the silica surface. This material was

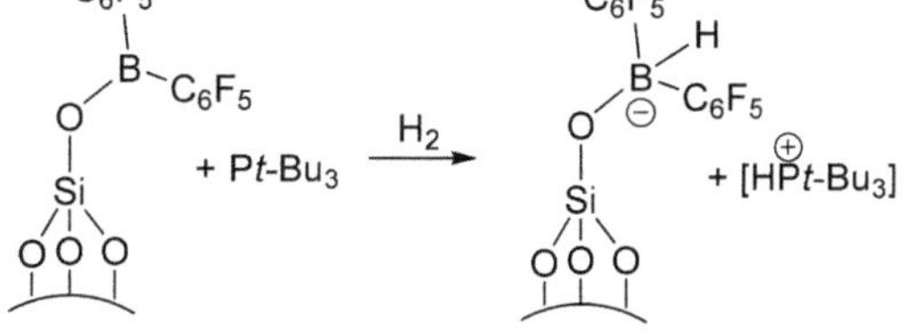

Scheme 6.41 Reaction of supported FLP with dihydrogen.

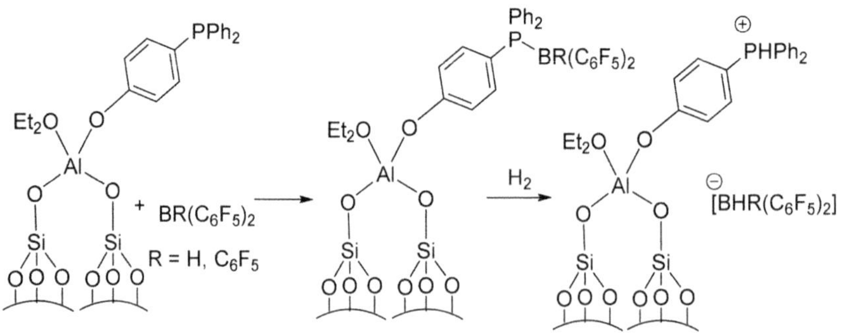

Scheme 6.42   Reaction of supported B–P adduct with dihydrogen.

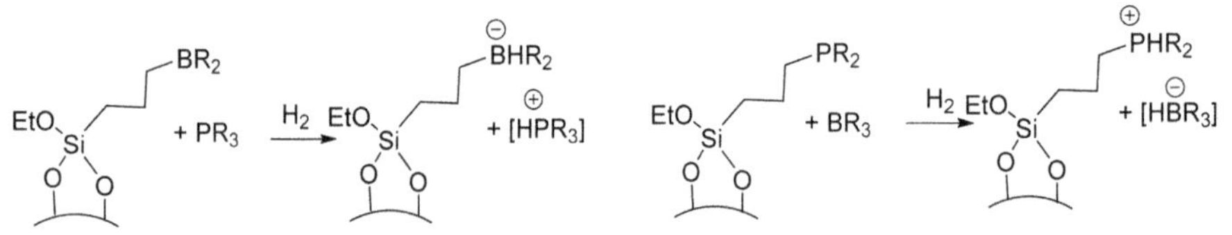

Scheme 6.43   Reactions of borane or phosphine functionalized silica with dihydrogen.

then treated with the boranes $B(C_6F_5)_3$ or $HB(C_6F_5)_2$, providing the silica support B–P adducts. Both materials activated dihydrogen and showed catalytic activity for the hydrogenation of alkynes. In the case of hexyne, these reductions were up to 87% *Z*-selective, while for other alkynes this selectivity was more moderate. These catalysts are thought to operate by dissociation of the supported B–P adduct. In addition, these catalysts proved recyclable with the addition of borane.

Another approach to supporting FLPs on silica has exploited allyl-functionalized silica nano-powders.[101] In this fashion borane and phosphine-derived silica nano-powders have been prepared. The addition of the complementary FLP partners to these materials resulted in supported FLPs. These materials were shown to activate dihydrogen (Scheme 6.43) and subsequently react with CO_2 to give HCOOH. The best of the supported catalysts proved to be the one derived from pendant PPh_2-groups in combination with BPh_3.

Another approach to a supported FLP has been the use of the solid polyamine derived from the reaction of *p*-xylylene diamine and 1,4-bis(bromomethyl)benzene. This, in combination with $B(C_6F_5)_3$, provided a semi-solid FLP which acted as a catalyst for the hydrogenation of $PhCH=C(CO_2Et)_2$.[102] In a conceptually related approach Thomas and coworkers[103] have prepared porous polymer networks derived from sterically encumbered triphenylphosphine precursors. These materials, in combination with $B(C_6F_5)_3$, yielded semi-immobilized FLPs, which cleaved dihydrogen at ambient temperature and low pressures.

6.14 Zeolite and MOF Derived FLPs

In 2015, Yoon and coworkers[104] considered the treatment of zeolite loaded with Pt nanoparticles with dihydrogen. Based on spectroscopy, these authors formulated the reactivity as an FLP generating a sodium hydride and protonated hydroxy oxygen atom. This species was confirmed to provide intra-zeolite reactivity as it proved to be a size-selective catalyst for the esterification of aldehydes, with acetaldehyde readily reacting whereas benzaldehyde was unreactive. On this basis, the authors speculated that this material could also be used in catalytic hydrogenation of organic compounds.

Another approach to the inclusion of catalytically active FLP sites into materials that has recently emerged is the use of metal-organic frameworks (MOFs). Computational studies in 2016[105] and 2020[106] explored the concept of the incorporation of BF_2 substituents on the organic linkers of the MOF UiO-66. Alternatively, a further computational study considered the generation of an FLP in the MOF by the removal of one of the organic linkers.[107] All of these studies have suggested the utility of these MOF-derived FLPs in CO_2 reduction chemistry.

In 2018, it was Ma *et al.*[108] that reported the first experimental work demonstrating the incorporation of a Lewis pair into a MOF framework. These authors exposed the MOF, MIL-101(Cr),[109] to a solution of 1,4-diazabicyclo[2.2.2]octane (DABCO), and subsequently $B(C_6F_5)_3$, resulting in the incorporation of the Lewis pair (DABCO)/$B(C_6F_5)_3$ into the MOF (Scheme 6.44). This material was employed as a hydrogenation catalyst for the size-selective reduction of imines. Similarly, this MOF-based catalyst also reduced alkylidene malonates. A subsequent study[110] by the same group showed the same system could be employed for the chemoselective hydrogenation of α,β-unsaturated organic compounds using moderate pressure of dihydrogen.

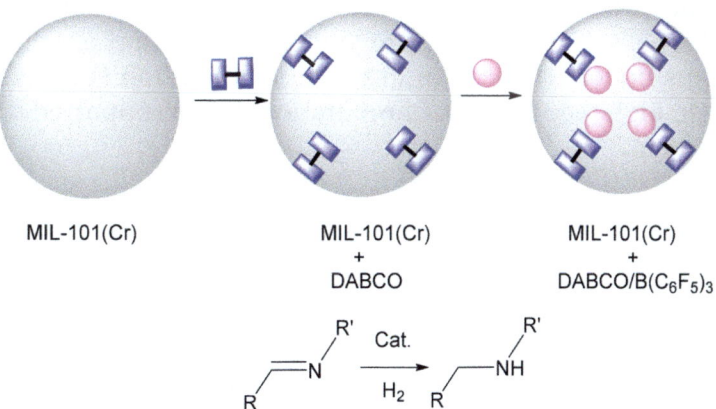

Scheme 6.44 Schematic representation of the inclusion of a Lewis pair into the MOF, MIL-101(Cr) and its use in hydrogenation catalysis.

A collaboration between the Dyson and Stylianou groups[111] demonstrated the use of porphyrin derived MOF, MOF-545, for the incorporation $B(C_6F_5)_3$. On exposure to CO_2 and dihydrogen, these materials affected the reduction of CO_2 to methoxide at a relatively low temperature and pressure. Treatment with an N-base prompted the release of the ammonium $[MeOB(C_6F_5)_3]$ allowing the recycling of the MOF.

Most recently, the classic MOF material $K_2Zn_3[Fe(CN)_6]_2$ was shown to exhibit catalytic hydrogenation activity *via* heterolytic cleavage of the dihydrogen. Computations supported its behavior as a Zn/N FLP.[112] This stable and recyclable catalyst was exploited to establish synthetic routes to 4-hydroxy-2-cyclopentenone and 2-cyclopentenone from biomass-derived furfuryl alcohol.

6.15　Heterogeneous Gold Catalysts

Other avenues to heterogeneous FLP systems have also emerged. The groups of Guo and Wang[113] showed that while a clean gold surface is unreactive with dihydrogen combination with an imine or nitrile in solution resulted in the hydrogenation of C–N multiple bonds (Scheme 6.45). These authors described this as FLP reactivity, where an Au Lewis acid and N-base donor activates dihydrogen in a biphasic FLP catalyst.

A similar FLP system has been derived from the secondary phosphine oxides in combination with gold nano-particles.[114] These materials were used in the chemoselective hydrogenation of acrolein and other α,β-unsaturated aldehydes. Similarly, gold nano-particles in combination with N-based bases also proved to be effective FLP catalysts for the hydrogenation of a series of internal and terminal alkynes[115] to the corresponding olefins. In the case of the internal olefins, the FLP reduction led to the *cis*-olefins.

In an interesting alternative to the above approaches to the use of gold nanoparticles, the pyrolysis of $Au(OAc)_3$ in the presence of 1,10-phenanthroline over TiO_2 furnished Au nanoparticles doped with nitrogen, carbon, and TiO_2.[116] This material proved to be a highly selective catalyst for the hydrogenation of alkynes to the corresponding olefins under relatively mild conditions (80–100 °C, 6–8 bar). This system also exhibited high recyclability. Computational data supported the notion of an Au/N FLP mechanism.

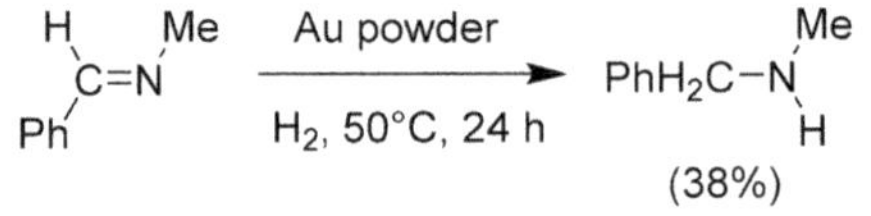

$$H \quad Me \qquad \text{Au powder} \qquad\qquad Me$$
$$\underset{Ph}{\diagup}C=N\diagup \xrightarrow[\text{H}_2,\ 50°\text{C},\ 24\ h]{} PhH_2C-N\underset{H}{\diagdown}$$
$$(38\%)$$

Scheme 6.45　FLP hydrogenation over a gold surface.

6.16 Metal-oxide FLP Catalysts

In 2015, Ozin and coworkers[117] recognized that hydroxylated indium oxide nanocrystals $(In_2O_3\ x(OH)_y)$ mediated the catalytic conversion of CO_2 to CO in the presence of dihydrogen *via* the reverse water gas shift reaction. In this case, the dihydrogen was activated between the Lewis basic hydroxide and Lewis acidic indium, prompting subsequent reaction with CO_2. This was also supported by *ab initio* molecular dynamics calculations.[118] A subsequent study[119] of the photochemistry of such indium oxide surfaces supported the notion that photoexcitation enhanced Lewis acidity and basicity, and thus improved catalytic activity. Further experimental and computational studies[120] showed that supported dihydrogen cleavage on the surface afforded a protonated hydroxide $(In–OH_2{}^+)$ and an indium hydride $(In–H^-)$. Adapting the above findings to the lesser-known rhombohedral polymorph of indium sesquioxide[121] significantly improved the catalytic activity, stability, and selectivity for the formation of MeOH. The gains in catalyst performance were attributed to the enhanced acidity and basicity of surface FLP in the rhombohedral form. More recently, Ozin's group[122,123] has engineered the isomorphic replacement of In(III) ions in In_2O_3 by Bi(III) ions. The resulting $Bi_xIn_{2-x}O_3$ materials were three orders of magnitude more photoactive for the reverse water gas shift converting CO_2 in the presence of dihydrogen to methanol and water.

Another approach to related heterogenous FLP catalysts involved the use of surface defects of porous nanorods of ceria.[124] This material was found to possess high concentrations of surface defects, providing Lewis acidic Ce(III) sites with neighboring basic oxygen atoms. While this heterogeneous FLP proved to be an efficient catalyst for the hydrogenation of alkenes and alkynes, the FLP mechanism was further supported in a detailed computational study.[125] Nano-rods of CeO_2 were much more active than CeO_2 nanocubes and nanopolyhedra,[126] in the hydrogenation catalysis of crotonaldehyde to crotyl alcohol. This was attributed to a higher degree of surface oxygen vacancies, thus affording an increase in the sites capable of FLP activation of dihydrogen. In a related study,[127] porous CeO_2 nano-rods were employed as an FLP catalyst for the activation of dihydrogen as well as the hydrogenation of alkenes and alkynes to alkanes.

Doping of ceria with metals provided reactive modifications. With nickel[128] the resulting nickel-doped ceria was found to generate oxygen vacancies and thus nickel served as a single-atom promoter. As a result, the nickel-doped ceria exhibited much higher activity for the hydrogenation of acetylene. In a closely related study,[129] gallium-doped ceria was studied. In contrast to nickel, the gallium dopant was found to act both as a promoter and as an active participant in the catalysis. In a very recent study,[130] as little as 2 mol% of the ruthenium doped ceria, Ru/CeO_2, acted as a catalyst for the hydrogenation of aromatic amines to the corresponding cyclohexylamine.

Other metal-oxide supports have also been probed. For example, dispersed gold and platinum on WO_x provided a catalyst for the selective hydrogenolysis of glycerol to 1,3-propanediol.[131] The dispersed gold is thought to facilitate the formation of surface FLP sites for the activation of dihydrogen. Yet another approach to heterogeneous interfacial FLPs was derived from cobalt borate $(CoBO_x)$ nano-sheets.[132] In these materials,

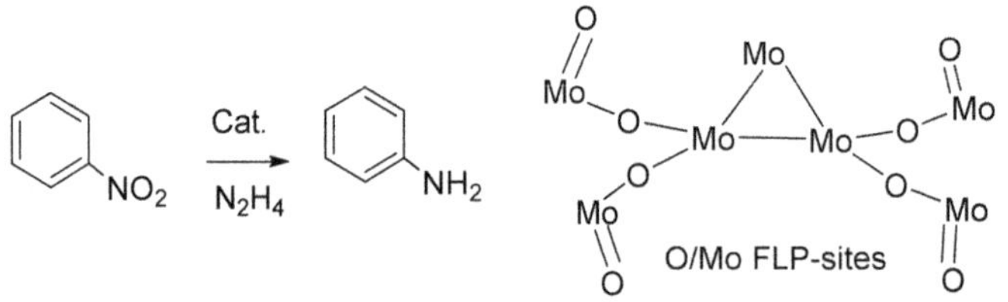

Scheme 6.46 Catalytic transfer hydrogenation using FLP-site in polyoxometalate type.

the surface hydroxyl on a Co atom acts as a basic site, while the adjacent interfacial cobalt ion is acidic in nature. This enabled the efficient activation of dihydrogen, thus permitting the hydrogenation of styrene.

More recently, a molecular polyoxometallate cluster $[(C_5H_5N)_3(Mo_{13}O_{32}(OH)_4)]$ (Scheme 6.43) was prepared and characterized.[133] Subsequent studies probed related polyoxometalate derivatives formulated as $H_4[Mo_{21}O_{54}(OH)_4py_6]$, $[Mo_{16}Zn_{0.5}Ge_{2.5}O_{47.5}(H_2O)_{0.5}py_6]$,[134] and $H[Sb_3Mo_{18}O_{55}(OH)_2py_3]$.[135] These species acted as an efficient hydrogenation catalyst for the reduction of nitroarenes to anilines using hydrazine hydrate as the hydrogen source. The authors proposed that this species acted as an FLP with one of the oxygen atoms of the polyoxometalate fragment acting as the base and one of the Mo centers of the trimetallic fragment acting as the acid, thus generating an O/M FLP (Scheme 6.46). These molecular species can thus be viewed as models for the metal-oxide surface FLP chemistry described above.

6.17   Other Solid-supported FLPs

Computational studies[136] have suggested the incorporation of a boron and nitrogen co-doped bilayer in graphene (BN-G) and graphene ribbon (BN-GR) would provide effective FLP catalysts for hydrogenation. In an experimental effort, N-doped reduced graphene oxide (N-RGO), was shown to be an effective hydrogenation catalyst for the reduction of ethylene and acetylene. The catalytic activity was attributed to the nitrogen sites in the graphene exhibiting FLP behavior.[137] In a conceptually related work, a Ni–N–C material was prepared by pyrolysis of Ni(phen)$_3$ dispersed on MgO at 600 °C after which the MgO was removed by acid etching.[138] This material proved to be an active catalyst for the hydrogenation of unsaturated substrates. Most importantly, these materials proved highly durable under very harsh conditions (245 °C, 60 bar dihydrogen, in the presence of aqueous tungstic acid). The activity of these materials was attributed to the proximity of nickel and N atoms, which act as FLP sites for the activation of dihydrogen.

DFT computations have also been used to propose two-dimensional FLP catalysts derived from a phosphorene monolayer doped with B or Al.[139] These species have been calculated to be effective hydrogenation catalysts for small unsaturated molecules, such as ketone, nitrile and ethylene. In a related notion, hybrid nanosheet materials derived from manganese oxides and cobalt phosphides[140] showed activity for the evolution of dihydrogen from alkaline solution. Computations indicated that the interface between Mn–O and Co–P acted as an FLP providing the driving force for water splitting.

6.18 Final Thoughts

In this chapter, we have discussed many areas of FLP-dihydrogen chemistry. Reductive FLP routes to radicals and the synthetic utility of transfer hydrogenations and dehydrogenation, expand the range of organic compounds available by FLP chemistry. The investigations of FLP reductions of CO and CO_2 chemistry, as well as the oxidation of dihydrogen, provide new opportunities for green energy sources. Extensions to main group hydrogenations and dinitrogen reductions are less well developed but certainly offer exciting opportunities to uncover new and important reactions. Moreover, the application of the concept of FLPs as it relates to enzymatic systems is both interesting and certainly demonstrates that this axiom of reactivity has reached well beyond main group chemistry.

The latter sections of this chapter have described solid-state and supported FLP catalysts. The incorporation of FLPs in polymers or MOFs are exciting new directions. Various approaches have also applied the concept of FLPs to heterogeneous catalysts. For example, the notion of surface FLPs as the sites of reactivity on metal-oxides further emphasizes a broadening application of the concept in catalysis.

In a relatively short period of 15 years, FLP-dihydrogen chemistry has developed and broadened dramatically from discovery to applications in a broad range of organic reductions, including highly selective, metal-free avenues to asymmetric hydrogenations. These developments have garnered much attention in the organic community. Meanwhile, the inorganic community has demonstrated that FLP-dihydrogen chemistry can be achieved without reliance on traditional Lewis acids and bases. Indeed, a range of main group species as well as transition, alkali, alkaline earth, and rare earth metals have been shown to participate in FLP reactivity. Beyond the traditional synthetic communities, the concept has also garnered much attention among those focused on theoretical and bioinorganic chemistry, as well as those developing heterogeneous catalysts. The uncovering of the power of combinations of electron-rich and electron-deficient centers in systems that avoid facile quenching has provided a new conceptual approach to developing unprecedented reactivity and illuminating new insight.

This text has focused on the reactivity of FLP systems with dihydrogen. It is important to note that simultaneously with the work described herein, the reactivity of FLPs with a broad array of other small molecules has also been probed, further increasing attention and broadening applications of FLP chemistry. For example, the development of FLP catalysts for polymerization,[139,141,142] the activation of C–H bonds by FLPs,[143] and the development of polymeric FLP sensors[144] are but three examples of areas where the concept has yielded new and seminal findings. Indeed, the original finding of the reaction of FLPs with dihydrogen has brought us a long way, and there is no doubt it will continue to stimulate the creativity of chemists.

References

1. X. Tao, G. Kehr, X. Wang, C. G. Daniliuc, S. Grimme and G. Erker, *Chem. - Eur. J.*, 2016, **22**, 9504–9507.
2. L. E. Longobardi, L. Liu, S. Grimme and D. W. Stephan, *J. Am. Chem. Soc.*, 2016, **138**, 2500–2503.
3. F. Calderazzo, C. Forte, F. Marchetti, G. Pampaloni and L. Pieretti, *Helv. Chim. Acta*, 2004, **87**, 781–789.
4. K. L. Bamford, L. E. Longobardi, L. Liu, S. Grimme and D. W. Stephan, *Dalton Trans.*, 2017, **46**, 5308–5319.

5. R. Noyori, *Angew. Chem., Int. Ed.*, 2002, **41**, 2008–2022.
6. T. Ohkuma, N. Utsumi, K. Tsutsumi, K. Murata, C. Sandoval and R. Noyori, *J. Am. Chem. Soc.*, 2006, **128**, 8724–8725.
7. N. Uematsu, A. Fujii, S. Hashiguchi, T. Ikariya and R. Noyori, *J. Am. Chem. Soc.*, 1996, **118**, 4916–4917.
8. F. Focante, P. Mercandelli, A. Sironi and L. Resconi, *Coord. Chem. Rev.*, 2006, **250**, 170–188.
9. G. Kehr, R. Roesmann, R. Fröhlich, C. Holst and G. Erker, *Eur. J. Inorg. Chem.*, 2001, 535–538.
10. J. M. Farrell, Z. M. Heiden and D. W. Stephan, *Organometallics*, 2011, **30**, 4497–4500.
11. P. Spies, S. Schwendemann, S. Lange, G. Kehr, R. Fröhlich and G. Erker, *Angew. Chem., Int. Ed.*, 2008, **47**, 7543–7546.
12. M. V. Mane, M. A. Rizvi and K. Vanka, *J. Org. Chem.*, 2015, **80**, 2081–2091.
13. A. F. G. Maier, S. Tussing, T. Schneider, U. Florke, Z. W. Qu, S. Grimme and J. Paradies, *Angew. Chem., Int. Ed.*, 2016, **55**, 12219–12223.
14. M. M. Guru, S. De, S. Dutta, D. Koley and B. Maji, *Chem. Sci.*, 2019, **10**, 7964–7974.
15. E. R. Clark and M. J. Ingleson, *Angew. Chem., Int. Ed.*, 2014, **53**, 11306–11309.
16. C. B. Caputo, L. J. Hounjet, R. Dobrovetsky and D. W. Stephan, *Science*, 2013, **341**, 1374–1377.
17. M. Perez, L. J. Hounjet, C. B. Caputo, R. Dobrovetsky and D. W. Stephan, *J. Am. Chem. Soc.*, 2013, **135**, 18308–18310.
18. M. Pérez, C. B. Caputo, R. Dobrovetsky and D. W. Stephan, *Proc. Natl. Acad. Sci. U. S. A.*, 2014, **111**, 10917–10921.
19. I. Chatterjee and M. Oestreich, *Angew. Chem., Int. Ed.*, 2015, **54**, 1965–1968.
20. W. Yuan, P. Orecchia and M. Oestreich, *Chem. Commun.*, 2017, **53**, 10390–10393.
21. I. Khan, B. G. Reed-Berendt, R. L. Melen and L. C. Morrill, *Angew. Chem., Int. Ed.*, 2018, **57**, 12356–12359.
22. A. C. Shaikh, J. M. Veleta, J. Moutet and T. L. Gianetti, *Chem. Sci.*, 2021, **12**, 4841–4849.
23. S. L. Li, G. Li, W. Meng and H. F. Du, *J. Am. Chem. Soc.*, 2016, **138**, 12956–12962.
24. S. Li, W. Meng and H. F. Du, *Org. Lett.*, 2017, **19**, 2604–2606.
25. W. Zhao, X. Feng, J. Yang and H. Du, *Tetrahedron Lett.*, 2019, **60**, 1193–1196.
26. A. J. M. Miller and J. E. Bercaw, *Chem. Commun.*, 2010, **46**, 1709–1711.
27. G. R. Whittell, E. I. Balmond, A. P. M. Robertson, S. K. Patra, M. F. Haddow and I. Manners, *Eur. J. Inorg. Chem.*, 2010, **2010**, 3967–3975.
28. C. Appelt, J. C. Slootweg, K. Lammertsma and W. Uhl, *Angew. Chem., Int. Ed.*, 2013, **52**, 4256–4259.
29. A. M. Chapman, M. F. Haddow and D. F. Wass, *J. Am. Chem. Soc.*, 2011, **133**, 8826–8829.
30. A. Robertson, G. Whittell, A. Staubitz, K. Lee, A. Lough and I. Manners, *Eur. J. Inorg. Chem.*, 2011, 5279–5287.
31. M. W. Lui, N. R. Paisley, R. McDonald, M. J. Ferguson and E. Rivard, *Chem. - Eur. J.*, 2016, **22**, 2134–2145.
32. Z. Mo, A. Rit, J. Campos, E. L. Kolychev and S. Aldridge, *J. Am. Chem. Soc.*, 2016, **138**, 3306–3309.
33. J. Possart and W. Uhl, *Organometallics*, 2018, **37**, 1314–1323.
34. M. Boudjelel, E. D. Sosa Carrizo, S. Mallet-Ladeira, S. Massou, K. Miqueu, G. Bouhadir and D. Bourissou, *ACS Catal.*, 2018, **8**, 4459–4464.
35. D. H. A. Boom, E. J. J. De Boed, E. Nicolas, M. Nieger, A. W. Ehlers, A. R. Jupp and J. C. Slootweg, *Z. Anorg. Allg. Chem.*, 2020, **646**, 586–592.
36. G. Ma, G. Song and Z. H. Li, *Chem. - Eur. J.*, 2018, **24**, 13238–13245.
37. I. Bhattacharjee, S. Bhunya and A. Paul, *Inorg. Chem.*, 2020, **59**, 1046–1056.
38. L. Zhang, T. Oishi, L. Gao, S. Hu, L. Yang, W. Li, S. Wu, R. Shang, Y. Yamamoto, S. Li, W. Wang and G. Zeng, *ChemPhysChem*, 2020, **21**, 2573–2578.
39. E. S. Osipova, E. S. Gulyaeva, E. I. Gutsul, V. A. Kirkina, A. A. Pavlov, Y. V. Nelyubina, A. Rossin, M. Peruzzini, L. M. Epstein, N. V. Belkova, O. A. Filippov and E. S. Shubina, *Chem. Sci.*, 2021, **12**, 3682–3692.
40. L. Wu, S. S. Chitnis, H. Jiao, V. T. Annibale and I. Manners, *J. Am. Chem. Soc.*, 2017, **139**, 16780–16790.
41. C. P. Sindlinger, F. S. W. Aicher, H. Schubert and L. Wesemann, *Angew. Chem., Int. Ed.*, 2017, **56**, 2198–2202.
42. Z. Mo, T. Szilvasi, Y. P. Zhou, S. Yao and M. Driess, *Angew. Chem., Int. Ed.*, 2017, **56**, 3699–3702.
43. B. D. Dombek, *Adv. Catal.*, 1983, **32**, 325–416.
44. P. M. Maitlis, *J. Organomet. Chem.*, 2004, **689**, 4366–4374.
45. A. J. M. Miller, J. A. Labinger and J. E. Bercaw, *J. Am. Chem. Soc.*, 2010, **132**, 3301–3303.
46. R. Dobrovetsky and D. W. Stephan, *J. Am. Chem. Soc.*, 2013, **135**, 4974–4977.
47. M. Sajid, G. Kehr, C. G. Daniliuc and G. Erker, *Angew. Chem., Int. Ed.*, 2014, **53**, 1118–1121.
48. Z. Jian, G. Kehr, C. G. Daniliuc, B. Wibbeling, T. Wiegand, M. Siedow, H. Eckert, M. Bursch, S. Grimme and G. Erker, *J. Am. Chem. Soc.*, 2017, **139**, 6474–6483.
49. M. Xu, Z.-W. Qu, S. Grimme and D. W. Stephan, *J. Am. Chem. Soc.*, 2021, **143**, 634–638.

50. C. M. Mömming, E. Otten, G. Kehr, R. Fröhlich, S. Grimme, D. W. Stephan and G. Erker, *Angew. Chem., Int. Ed.*, 2009, **48**, 6643–6646.
51. A. E. Ashley, A. L. Thompson and D. O'Hare, *Angew. Chem., Int. Ed.*, 2009, **48**, 9839–9843.
52. A. L. Travis, S. C. Binding, H. Zaher, T. A. Q. Arnold, J. C. Buffet and D. O'Hare, *Dalton Trans.*, 2013, **42**, 2431–2437.
53. S. D. Tran, T. A. Tronic, W. Kaminsky, D. M. Heinekey and J. M. Mayer, *Inorg. Chim. Acta*, 2011, **369**, 126–132.
54. L. Zhao, G. Lu, F. Huang and Z. Wang, *Dalton Trans.*, 2012, **41**, 4674–4684.
55. B. Jiang, Q. Zhang and L. Dang, *Org. Chem. Front.*, 2018, **5**, 1905–1915.
56. M. Ghara and P. K. Chattaraj, *Struct. Chem.*, 2019, **30**, 1067–1077.
57. Z. P. Lu, Y. W. Wang, J. Liu, Y. J. Lin, Z. H. Li and H. D. Wang, *Organometallics*, 2013, **32**, 6753–6758.
58. D. W. Stephan and G. Erker, *Chem. Sci.*, 2014, **5**, 2625–2641.
59. M. A. Courtemanche, A. P. Pulis, É. Rochette, M. A. Légaré, D. W. Stephan and F. G. Fontaine, *Chem. Commun.*, 2015, **51**, 9797–9800.
60. Y. F. Jiang, O. Blacque, T. Fox and H. Berke, *J. Am. Chem. Soc.*, 2013, **135**, 7751–7760.
61. C. M. Zall, J. C. Linehan and A. M. Appel, *ACS Catal.*, 2015, **5**, 5301–5305.
62. C. M. Zall, J. C. Linehan and A. M. Appel, *J. Am. Chem. Soc.*, 2016, **138**, 9968–9977.
63. R. Watari, S. Kuwata and Y. Kayaki, *Chem. Lett.*, 2020, **49**, 252.
64. K. Takeuchi, Y. Tanaka, I. Tanigawa, F. Ozawa and J.-C. Choi, *Dalton Trans.*, 2020, **49**, 3630–3637.
65. H. Ge, Y. Jing and X. Yang, *Inorg. Chem.*, 2016, **55**, 12179–12184.
66. X. Chen, Y. Jing and X. Yang, *Chem. - Eur. J.*, 2016, **22**, 8897–8902.
67. E. A. Romero, T. X. Zhao, R. Nakano, X. B. Hu, Y. T. Wu, R. Jazzar and G. Bertrand, *Nat. Catal.*, 2018, **1**, 743–747.
68. V. B. Saptal, G. Juneja and B. M. Bhanage, *New J. Chem.*, 2018, **42**, 15847–15851.
69. E. J. Lawrence, V. S. Oganesyan, D. L. Hughes, A. E. Ashley and G. G. Wildgoose, *J. Am. Chem. Soc.*, 2014, **136**, 6031–6036.
70. E. J. Lawrence, R. J. Blagg, D. L. Hughes, A. E. Ashley and G. G. Wildgoose, *Chem. - Eur. J.*, 2015, **21**, 900–906.
71. E. J. Lawrence, T. J. Herrington, A. E. Ashley and G. G. Wildgoose, *Angew. Chem., Int. Ed.*, 2014, **53**, 9922–9925.
72. E. J. Lawrence, E. R. Clark, L. D. Curless, J. M. Courtney, R. J. Blagg, M. J. Ingleson and G. G. Wildgoose, *Chem. Sci.*, 2016, 7, 2537–2543.
73. E. L. Bennett, E. J. Lawrence, R. J. Blagg, A. S. Mullen, F. Macmillan, A. W. Ehlers, D. J. Scott, J. S. Sapsford, A. E. Ashley, G. G. Wildgoose and J. C. Slootweg, *Angew. Chem., Int. Ed.*, 2019, **58**, 8362–8366.
74. S. J. Geier and D. W. Stephan, *Chem. Commun.*, 2010, **46**, 1026–1028.
75. S. J. Geier, M. A. Dureen, E. Y. Ouyang and D. W. Stephan, *Chem. - Eur. J.*, 2010, **16**, 988–993.
76. R. Dobrovetsky, K. Takeuchi and D. W. Stephan, *Chem. Commun.*, 2015, **51**, 2396–2398.
77. A. J. Stepen, M. Bursch, S. Grimme, D. W. Stephan and J. H. Paradies, *Angew. Chem., Int. Ed.*, 2018, **57**, 15253–15256.
78. H. Zhu, Z.-W. Qu and S. Grimme, *Chem. - Eur. J.*, 2019, **25**, 4670–4672.
79. K. C. Janda, L. S. Bernstein, J. M. Steed, S. E. Novick and W. Klemperer, *J. Am. Chem. Soc.*, 1978, **100**, 8074.
80. R. Appel and R. Schöllhorn, *Angew. Chem., Int. Ed.*, 1964, **3**, 805.
81. N. Holzmann, D. Dange, C. Jones and G. Frenking, *Angew. Chem., Int. Ed.*, 2013, **52**, 3004–3008.
82. T. Itoh, Y. Nakata, K. Hirai and H. Tomioka, *J. Am. Chem. Soc.*, 2006, **128**, 957–967.
83. R. C. Neu and D. W. Stephan, *Organometallics*, 2012, **31**, 46–49.
84. C. N. Tang, Q. M. Liang, A. R. Jupp, T. C. Johnstone, R. C. Neu, D. T. Song, S. Grimme and D. W. Stephan, *Angew. Chem., Int. Ed.*, 2017, **56**, 16588–16592.
85. L. L. Cao, J. Zhou, Z. W. Qu and D. W. Stephan, *Angew. Chem., Int. Ed.*, 2019, **58**, 18487–18491.
86. M.-A. Légaré, G. Bélanger-Chabot, R. D. Dewhurst, E. Welz, I. Krummenacher, B. Engels and H. Braunschweig, *Science*, 2018, **359**, 896–900.
87. M.-A. Légaré, M. Rang, G. Bélanger-Chabot, J. I. Schweizer, I. Krummenacher, R. Bertermann, M. Arrowsmith, M. C. Holthausen and H. Braunschweig, *Science*, 2019, **363**, 1329–1332.
88. J. B. Geri, J. P. Shanahan and N. K. Szymczak, *J. Am. Chem. Soc.*, 2017, **139**, 5952–5956.
89. A. Simonneau, R. Turrel, L. Vendier and M. Etienne, *Angew. Chem., Int. Ed.*, 2017, **56**, 12268–12272.
90. M. Tamizmani and C. Sivasankar, *Eur. J. Inorg. Chem.*, 2017, **2017**, 4239–4245.
91. T. C. Jeyakumar, S. Baskaran and C. Sivasankar, *J. Chem. Sci.*, 2018, **130**, 57.
92. L. R. Doyle, A. J. Wooles and S. T. Liddle, *Angew. Chem., Int. Ed.*, 2019, **58**, 6674–6677.

93. L. Chatelain, E. Louyriac, I. Douair, E. Lu, F. Tuna, A. J. Wooles, B. M. Gardner, L. Maron and S. T. Liddle, *Nat. Commun.*, 2020, **11**, 337.
94. S. Shima, O. Pilak, S. Vogt, M. Schick, M. S. Stagni, W. Meyer-Klaucke, E. Warkentin, R. K. Thauer and U. Ermler, *Science*, 2008, **321**, 572–575.
95. K. F. Kalz, A. Brinkmeier, S. Dechert, R. A. Mata and F. Meyer, *J. Am. Chem. Soc.*, 2014, **136**, 16626–16634.
96. W. Lubitz, H. Ogata, O. Rüdiger and E. Reijerse, *Chem. Rev.*, 2014, **114**, 4081–4148.
97. G. Erker, L. Wang, G. Kehr, C. G. Daniliuc, M. Brinkkötter, T. Wiegand, A.-L. Wübker, H. Eckert, L. Liu, J. G. Brandenburg and S. Grimme, *Chem. Sci.*, 2018, **9**, 4859–4865.
98. M. E. Bowden, B. Ginovska, M. O. Jones, A. J. Karkamkar, A. J. Ramirez-Cuesta, L. L. Daemen, G. K. Schenter, S. A. Miller, T. Repo, K. Chernichenko, N. Leick, M. B. Martinez and T. Autrey, *Inorg. Chem.*, 2020, **59**, 15295–15301.
99. J. Y. Xing, J. C. Buffet, N. H. Rees, P. Norby and D. O'Hare, *Chem. Commun.*, 2016, **52**, 10478–10481.
100. K. C. Szeto, W. Sahyoun, N. Merle, J. L. Castelbou, N. Popoff, F. Lefebvre, J. Raynaud, C. Godard, C. Claver, L. Delevoye, R. M. Gauvin and M. Taoufik, *Catal. Sci. Technol.*, 2016, **6**, 882–889.
101. K. Mentoor, L. Twigge, J. W. H. Niemantsverdriet, J. C. Swarts and E. Erasmus, *Inorg. Chem.*, 2021, **60**, 55–69.
102. A. Willms, H. Schumacher, T. Tabassum, L. Qi, S. L. Scott, P. J. C. Hausoul and M. Rose, *ChemCatChem*, 2018, **10**, 1835–1843.
103. M. Trunk, J. F. Teichert and A. Thomas, *J. Am. Chem. Soc.*, 2017, **139**, 3615–3618.
104. H. Lee, Y. N. Choi, D. W. Lim, M. M. Rahman, Y. I. Kim, I. H. Cho, H. W. Kang, J. H. Seo, C. Jeon and K. B. Yoon, *Angew. Chem., Int. Ed.*, 2015, **53**, 13080–13084.
105. J. Y. Ye and J. K. Johnson, *Catal. Sci. Technol.*, 2016, **6**, 8392–8405.
106. M. Heshmat, *J. Phys. Chem. C*, 2020, **124**, 10951–10960.
107. K. Yang and J. Jiang, *J. Mater. Chem. A*, 2020, **8**, 22802–22815.
108. Z. Niu, W. D. C. Bhagya-Gunatilleke, Q. Sun, P. C. Lan, J. Perman, J.-G. Ma, Y. Cheng, B. Aguila and S. Ma, *Chem*, 2018, 2587–2599.
109. L. Bromberg, Y. Diao, H. Wu, S. A. Speakman and T. A. Hatton, *Chem. Mater.*, 2012, **24**, 1664–1675.
110. Z. Niu, W. Zhang, P. C. Lan, B. Aguila and S. Ma, *Angew. Chem., Int. Ed.*, 2019, **58**, 7420–7424.
111. S. Shyshkanov, T. N. Nguyen, A. Chidambaram, K. C. Stylianou and P. J. Dyson, *Chem. Commun.*, 2019, **55**, 10964–10967.
112. X. Li, Q. Deng, L. Yu, R. Gao, Z. Tong, C. Lu, J. Wang, Z. Zeng, J.-J. Zou and S. Deng, *Green Chem.*, 2020, **22**, 2549–2557.
113. G. Lu, P. Zhang, D. Sun, L. Wang, K. Zhou, Z.-X. Wang and G.-C. Guo, *Chem. Sci.*, 2014, **5**, 1082.
114. N. Almora-Barrios, I. Cano, P. W. N. M. van Leeuwen and N. Lopez, *ACS Catal.*, 2017, **7**, 3949–3954.
115. J. L. Fiorio, N. Lopez and L. M. Rossi, *ACS Catal.*, 2017, **7**, 2973–2980.
116. J. L. Fiorio, R. V. Goncalves, E. Teixeira-Neto, M. A. Ortuno, N. Lopez and L. M. Rossi, *ACS Catal.*, 2018, **8**, 3516–3524.
117. K. K. Ghuman, T. E. Wood, L. B. Hoch, C. A. Mims, G. A. Ozin and C. V. Singh, *Phys. Chem. Chem. Phys.*, 2015, **17**, 14623–14635.
118. M. Ghoussoub, S. Yadav, K. K. Ghuman, G. A. Ozin and C. V. Singh, *ACS Catal.*, 2016, **6**, 7109–7117.
119. K. K. Ghuman, L. B. Hoch, P. Szymanski, J. Y. Loh, N. P. Kherani, M. A. El-Sayed, G. A. Ozin and C. V. Singh, *J. Am. Chem. Soc.*, 2016, **138**, 1206–1214.
120. L. Wang, T. Yan, R. Song, W. Sun, Y. Dong, J. Guo, Z. Zhang, X. Wang and G. A. Ozin, *Angew. Chem., Int. Ed.*, 2019, **58**, 9501–9505.
121. T. Yan, L. Wang, Y. Liang, M. Makaremi, T. E. Wood, Y. Dai, B. Huang, A. A. Jelle, Y. Dong and G. A. Ozin, *Nat. Commun.*, 2019, **10**, 1–10.
122. T. Yan, N. Li, L. Wang, W. Ran, P. N. Duchesne, L. Wan, N. T. Nguyen, L. Wang, M. Xia and G. A. Ozin, *Nat. Commun.*, 2020, **11**, 6095.
123. Y. Dong, K. K. Ghuman, R. Popescu, P. N. Duchesne, W. Zhou, J. Y. Y. Loh, A. A. Jelle, J. Jia, D. Wang, X. Mu, C. Kuebel, L. Wang, L. He, M. Ghoussoub, Q. Wang, T. E. Wood, L. M. Reyes, P. Zhang, N. P. Kherani, C. V. Singh and G. A. Ozin, *Adv. Sci.*, 2018, **5**, 1700732.
124. S. Zhang, Z.-Q. Huang, Y. Ma, W. Gao, J. Li, F. Cao, L. Li, C.-R. Chang and Y. Qu, *Nat. Commun.*, 2017, **8**, 15266.
125. Z.-Q. Huang, L.-P. Liu, S. Qi, S. Zhang, Y. Qu and C.-R. Chang, *ACS Catal.*, 2018, **8**, 546–554.
126. Z. Zhang, Z.-Q. Wang, Z. Li, W.-B. Zheng, L. Fan, J. Zhang, Y.-M. Hu, M.-F. Luo, X.-P. Wu, X.-Q. Gong, W. Huang and J.-Q. Lu, *ACS Catal.*, 2020, **10**, 14560–14566.
127. S. Zhang, M. Zhang and Y. Qu, *Acta Phys.-Chim. Sin.*, 2020, **36**, 1911050.

128. C. Riley, S. Zhou, D. Kunwar, A. De La Riva, E. Peterson, R. Payne, L. Gao, S. Lin, H. Guo and A. Datye, *J. Am. Chem. Soc.*, 2018, **140**, 12964–12973.
129. S. Zhou, L. Gao, F. Wei, S. Lin and H. Guo, *J. Catal.*, 2019, **375**, 410–418.
130. Y. Cao, H. Zheng, G. Zhu, H. Wu and L. He, *Chin. Chem. Lett.*, 2021, **32**, 770–774.
131. X. Zhao, J. Wang, M. Yang, N. Lei, L. Li, B. Hou, S. Miao, X. Pan, A. Wang and T. Zhang, *ChemSusChem*, 2017, **10**, 819–824.
132. S. Zhang, Z.-Q. Huang, X. Chen, J. Gan, X. Duan, B. Yang, C.-R. Chang and Y. Qu, *J. Catal.*, 2019, **372**, 142–150.
133. B. Luo, R. Sang, L. Lin and L. Xu, *Catal. Sci. Technol.*, 2019, **9**, 65–69.
134. B. Luo and L. Xu, *Dalton Trans.*, 2019, **48**, 6892–6898.
135. F. Yu and L. Xu, *Dalton Trans.*, 2019, **48**, 17445–17450.
136. X. Sun, B. Li, T. Liu, J. Song and D. S. Su, *Phys. Chem. Chem. Phys.*, 2016, **18**, 11120–11124.
137. A. A. Abakumov, I. B. Bychko, A. S. Nikolenko and P. E. Strizhak, *Theor. Exp. Chem.*, 2018, **54**, 218–224.
138. W. Liu, Y. Chen, H. Qi, L. Zhang, W. Yan, X. Liu, X. Yang, S. Miao, W. Wang, C. Liu, A. Wang, J. Li and T. Zhang, *Angew. Chem., Int. Ed.*, 2018, **57**, 7071–7075.
139. J. Zhao, X. Liu and Z. Chen, *ACS Catal.*, 2017, **7**, 766–771.
140. D. Zhou, Z. Wang, X. Long, Y. An, H. Lin, Z. Xing, M. Ma and S. Yang, *J. Mater. Chem. A*, 2019, **7**, 22530–22538.
141. Q. Wang, W. Zhao, S. Zhang, J. He, Y. Zhang and E. Y.-X. Chen, *ACS Catal.*, 2018, **8**, 3571–3578.
142. Y. T. Zhang, G. M. Miyake and E. Y. X. Chen, *Angew. Chem., Int. Ed.*, 2010, **49**, 10158–10162.
143. M. A. Legare, M. A. Courtemanche, E. Rochette and F. G. Fontaine, *Science*, 2015, **349**, 513–516.
144. M. Wang, F. Nudelman, R. R. Matthes and M. P. Shaver, *J. Am. Chem. Soc.*, 2017, **139**, 14232–14236.

Subject Index